Textual and contextual data analysis

Multidimensional statistical analysis of textual data is a powerful technique that enables researchers to uncover deeper insights into the context and meaning of documents. This book addresses the challenge of jointly analyzing textual and contextual data, presenting rigorous theoretical foundations alongside practical methodologies. By incorporating metadata and contextual information, readers can extract richer, more nuanced information from textual corpora, making this book an essential resource for statisticians, data scientists, and linguistics experts.

The book explores a wide range of textual data, from open-ended survey responses and political speeches to legal texts, literary works, and technical reports. It also examines the diverse contextual variables that shape these texts, such as sociodemographic characteristics, chronology, political affiliations, and external influences. Through real-world examples, readers will learn how to apply exploratory multivariate statistical methods to compare, characterize, and reveal the underlying structure of textual data. Each chapter builds on the previous one, offering a systematic approach to encoding, analyzing, and visualizing textual and contextual data. Topics include machine learning methods like latent semantic analysis and correspondence analysis, clustering techniques, restricted clustering defined by contextual data, and advanced visualization tools. The book also introduces methodologies for analyzing multilingual corpora and isolated texts, emphasizing the importance of discourse strategies and thematic contrasts.

This book is not only a guide to advanced statistical methods but also a practical toolkit for researchers working with diverse corpora. Whether analyzing legal databases, sensory evaluations, or political speeches, readers will find robust techniques to uncover patterns, relationships, and strategies within their data. By combining textual and contextual analysis, this book empowers readers to make meaningful comparisons and draw actionable conclusions.

KEY FEATURES:

- Comprehensive coverage of methods for jointly analyzing textual and contextual data.
- Practical applications to diverse corpora, including legal texts, political speeches, and sensory evaluations.
- Systematic comparison of machine learning methods like latent semantic analysis and correspondence analysis.
- Advanced visualization techniques, including interactive, 3D, and animated graphics.
- Methodologies for analyzing multilingual corpora and isolated texts, with a focus on discourse strategies.

CHAPMAN & HALL/CRC DATA SCIENCE SERIES

Reflecting the interdisciplinary nature of the field, this book series brings together researchers, practitioners, and instructors from statistics, computer science, machine learning, and analytics. The series will publish cutting-edge research, industry applications, and textbooks in data science.

The inclusion of concrete examples, applications, and methods is highly encouraged. The scope of the series includes titles in the areas of machine learning, pattern recognition, predictive analytics, business analytics, Big Data, visualization, programming, software, learning analytics, data wrangling, interactive graphics, and reproducible research.

Recently Published Titles

Research Software Engineering
A Guide to the Open Source Ecosystem
Matthias Bannert

The Data Preparation Journey
Finding Your Way With R
Martin Hugh Monkman

Getting (more out of) Graphics
Practice and Principles of Data Visualisation
Antony Unwin

Introduction to Data Science
Data Wrangling and Visualization with R Second Edition
Rafael A. Irizarry

Data Science
A First Introduction with Python
Tiffany Timbers, Trevor Campbell, Melissa Lee, Joel Ostblom and Lindsey Heagy

Mathematical Engineering of Deep Learning
Benoit Liquet, Sarat Moka, and Yoni Nazarathy

Introduction to Classifier Performance Analysis with R
Sutaip L.C. Saw

Predictive Modelling for Football Analytics
Leonardo Egidi, Karlis Dimitris, and Ioannis Ntzoufras

Deep-Learning-Assisted Statistical Methods with Examples in R
Tianyu Zhan

Test-Driven Data Analysis
Nicholas J. Radcliffe

What's The Question?
Deciding What You Really Want to Know
David J. Hand

For more information about this series, please visit: https://www.routledge.com/Chapman--HallCRC-Data-Science-Series/book-series/CHDSS

Textual and contextual data analysis

A Multivariate Statistical Approach using R

Mónica Bécue-Bertaut
and Ramón Alvarez-Esteban

CRC Press
Taylor & Francis Group
Boca Raton London New York

CRC Press is an imprint of the
Taylor & Francis Group, an **informa** business
A CHAPMAN & HALL BOOK

Designed Cover Image: Mónica Bécue-Bertaut and Ramón Alvarez-Esteban

First edition published 2026
by CRC Press
2385 NW Executive Center Drive, Suite 320, Boca Raton FL 33431

and by CRC Press
4 Park Square, Milton Park, Abingdon, Oxon, OX14 4RN

CRC Press is an imprint of Taylor & Francis Group, LLC

ISBN: 978-1-032-50226-7 (hbk)
ISBN: 978-1-032-50227-4 (pbk)
ISBN: 978-1-003-39747-2 (ebk)

DOI: 10.1201/9781003397472

Typeset in CMR10 Roman
by KnowledgeWorks Global Ltd.

Publisher's Note: This book has been prepared from camera-ready copy provided by the author

Contents

Preface

What we intend to present

Multidimensional statistical analysis of textual data is a widely practiced technique. This data usually includes metadata or contextual information. Incorporating this information into the analysis provides insight into the documents' context and enables the extraction of more information from them. This book addresses the problem of jointly analyzing textual and contextual data. It presents recent methods that facilitate this process. The theoretical aspects are approached rigorously. Numerous real-life examples are analyzed to illustrate the methodology and analysis strategy.

Textual data can include open-ended survey responses, political speeches, legal texts from extensive databases, expert reports, literary texts, and individual texts whose composition we intend to analyze. Studies in sensory analysis, in which participants provide free-text evaluations in addition to numerical ratings, can also be addressed.

The contextual data is also highly varied. Answers to open-ended questions are influenced by respondents' sociodemographic characteristics. Political speeches are enriched by considering the dates on which they were delivered and the political parties involved. A corpus of Supreme Court judicial sentences includes information such as the year, the names and political tendencies of the judges, the laws cited, the jurisprudence used, and the verdict. When dealing with literary corpora, the year of publication is important. Technical corpora, such as those documenting incidents at nuclear power plants, contain quantitative and qualitative information about these incidents. When studying a corpus over time—for example, the successive political speeches of a single political figure or a single closing speech delivered in court—chronology is an important contextual variable.

Our experience with diverse users has shown us that textual data can be extensive, as in the case of legal databases, or limited, as in sensory analysis, where the corpus comprises only a few dozen brief product descriptions. This data may be accompanied by contextual information comprising a wide range of variables.

This book showcases the robustness and adaptability of these methods by demonstrating their successful application to diverse corpora. The starting point is always a multiple table combining one or more textual tables (documents$\times$words) with one or more qualitative or quantitative contextual tables (documents$\times$variables). Depending on the nature of the problem and the data, these tables—and thus the textual and contextual data—play a symmetrical or nonsymmetrical role.

We advocate using exploratory multivariate statistical methods, which are primarily comparative tools. Effective characterization is only possible through comparison, and meaningful comparisons cannot be made from isolated elements. Our goal is to compare the vocabulary of texts or groups of texts to identify shared or specific topics. However, we also prioritize revealing the underlying structure of the corpus to understand its purpose. We aim to reveal the discursive strategies employed by speakers, as well as the sociodemographic characteristics shaping detected thematic contrasts. For this reason, we refer to this approach as a structural analysis of textual data.

Book content

Chapter 1 begins with a review of the fundamentals of encoding textual and contextual data. Then, it emphasizes the importance of contextual data and clarifies the roles these two types of data can play. It also reviews and briefly discusses some classical unidimensional methods.

Chapter 2 compares two competing text-oriented machine learning methods: latent semantic analysis (LSA) and correspondence analysis (CA). Both methods use singular value decomposition (SVD) to organize documents and words in a lower-dimensional space. In this respect, the two methods appear similar. However, their differences influence their usage, which is why we deemed a systematic comparison important.

Chapter 3 recalls and compares the classical clustering methods as applied to textual data. The outputs are analyzed from the perspectives of internal, relative, and external validation in order to compare the clustering results. A review of the main cluster analysis families is carried out, especially density methods such as DBSCAN.

Chapter 4 introduces restricted clustering, which is defined by contextual data, and includes the link-not-link, size, and contiguity. The Guttman effect on the first factorial plane is analyzed, as are techniques for minimizing its impact. Particular attention is given to presenting chronological contiguity and explaining local stopping rules.

Chapter 5 deals with the visualization and mapping of textual data. The classical two-dimensional graphs are enhanced with tools imported from spatial statistics and geostatistics. Representations are approached from two perspectives: summarizing the main characteristics of the phenomenon under study and extracting structures as a research tool. There are a variety of options for creating interactive, 3D, and animated graphics. This is a valuable and original contribution to visualization, an area that is becoming increasingly important in textual analysis.

Chapter 6 presents a methodology for equally analyzing textual and contextual data. One example is a questionnaire that uses open- and closed-ended questions to address a societal issue. Another interesting example could be a scoring system that combines numerical scores and evaluative comments.

Chapter 7 examines contextual data that plays an explanatory role in the analysis of textual data. The proposed methodology employs projected multidimensional methods closely related to regression methods. These methods treat the text being explained as multidimensional. In other words, the focus is on configurations of words rather than individual words.

Chapter 8 focuses on the study of the structure and organization of an isolated text. The main objective is to reveal the discourse strategy employed. This topic is particularly relevant in fields such as law and politics, where persuasive speech is essential.

Chapter 9 discusses applying multivariate statistical methods to multilingual corpora. We present a methodology that enables the simultaneous analysis of open-ended responses across languages. This methodology examines the relationship between word choice and the factors that influence it.

Readers of the book

This book compiles advanced methods that were previously only available in scattered articles. It will be of interest to a diverse audience, including statisticians, data scientists, and linguistics experts—in other words, anyone who needs to extract information from texts.

The R packages

The examples in the book use various packages from the open-source R software (https://cran.r-project.org), particularly our own package, **Xplortext** (Bécue-Bertaut et al., 2024). The corresponding databases and R scripts for the book examples can be found on the website https://xplortext.unileon.es. These resources allow readers to easily become familiar with the methods described in the book by reproducing the examples.

Acknowledgments

First and foremost, we would like to acknowledge the invaluable assistance and unwavering support provided by David Grubbs, the editor of this book at Taylor & Francis. We would also like to thank Professor Elias Rizkallah (UQAM, Montreal, Canada) for reviewing the structure and content of our book proposal. Chapters 6 and 7 are the result of our long-term collaboration with Jérôme Pagès. We are also indebted to B. Kostov (UPC, Barcelona, Spain) for his contributions, especially his work designing computer tools for his thesis. We greatly appreciated our discussions with Guilhem Naro (UPF, Barcelona, Spain). He helped us understand the discursive strategy used by the author of a rhetorical text. We are extremely grateful to M.C. Blanc and C. Álvarez for proofreading the English version. Finally, we would like to thank our families for their support and encouragement.

About the Authors

Dr. Mónica Bécue-Bertaut taught statistics and data science at the Universitat Politènica de Catalunya and offered numerous guest lectures on textual data science in different countries. She has published several books and chapters on this topic, and she has helped design software related to textual data science, including SPAD. T and the R package Xplortext. She is an elected fellow of the International Statistical Institute and a Chevalier des Palmes Académiques, a distinction bestowed by the French government.

Dr. Ramón Alvarez-Esteban is an associate professor at the University of León (Spain), where he teaches multivariate data analysis and R. His research interests include textual data analysis, climate change models, and integrated statistical and geospatial techniques. He is an author and the maintainer of the Xplortext R package (Statistical Analysis of Textual Data), which has been available on the CRAN website since 2017.

List of Figures

List of Tables

1

Consideration of additional information called contextual data

1.1 Textual data

1.1.1 What is meant by a corpus?

A corpus is a closed set of texts and/or data intended for a specific objective. The decisions involved in its design are guided by the user's objectives. Frequently, different fields use different languages (field-specific corpora). Consequently, the analysis of documents from a unique field may yield better results than the analysis of generic corpora or corpora originating from several fields. A corpus may already exist prior to the current study, such as when a collection of abstracts is downloaded from a scientific database to track the development of research in a given field. In other cases, it is created for the study itself. This is often the case when dealing with responses to an open-ended question or a series of interviews. The corpus constitution step is fundamental, as it determines the quality of the study. Precise hypotheses guide the user through the entire process. However, this exceeds the scope of this book, so we will limit ourselves to providing a few comments applied to several practical examples.

The non-textual parts are the classical statistical variables that characterize the documents, such as the gender of their authors. These variables are called contextual variables.

The statistical approach we are pursuing involves transforming the corpus (textual and non-textual parts) into cross-tabulations. For this purpose, the textual part of the corpus is divided into more or less fine fragments called texts or documents. The division may be *a priori*, as in the case of open-ended responses. It may be artificial, as in the case of a lengthy single text whose temporal structure we intend to identify, which requires dividing the text into parts or documents.

1.1.2 Types of corpora

The corpora used to apply the methods in this book are varied in type. It is common to distinguish between two types of corpora: those consisting of

DOI: 10.1201/9781003397472-1

a large number of short texts and those consisting of a small or moderate number of longer texts. In the first category, we find a corpus of responses to open-ended questions, typically enriched with a large set of contextual data (see Chapter 3 for an example).

In the second category, for example, we have a collection of speeches by successive politicians. In this case, the contextual data is typically limited to the publication date, the political party, and the politician's name (see Chapter 4).

Nevertheless, we can be led to analyze corpora consisting of a small number of very short texts, as in the case of tasting data, where a reduced number of panelists describe, in their own words, a small number of products. In this case, a huge amount of contextual data is available, making it possible to take advantage of the free comments (see Chapter 2). In other cases, such as the analysis of a collection of legal reasoning extracted from a large legal database, we face a large number of rather long texts. It is possible to analyze a corpus consisting of a single text that has been divided into parts for the purpose of analysis. Analyzing how the vocabulary evolves throughout these parts can help us better understand the structure of the text (see Chapter 8).

1.2 Contextual data

Contextual data play an essential role in exploring and analyzing textual data. Also known as metadata, contextual data describe documents, their authors, and the context in which the documents originated. In the case of open-ended survey questions, contextual data consist of closed-ended responses ranging from socioeconomic variables to opinions and practices.

In general, contextual data refer to the surrounding information that helps one fully understand the meaning of a text from a linguistic standpoint. This also explains why users select different words, linking word choices to socioeconomic characteristics, opinions, and practices. It is particularly useful for answering questions such as *who*, *what*, *where*, *when*, *how*, and *why*. Some contextual data are dynamic, meaning they can change over time. This type of data can provide contextual information—typically encoded as quantitative or qualitative statistical variables—that improves the interpretation of primary data. To identify the narrative structure of an extensive text, i.e., its narrative strategy, one must use contextual data to establish the chronological order of its parts (see example described in Section 1.7.4 and analyzed in Chapter 8). It is possible to count the frequency of a word within a text and compare it with the frequency of the other words. However, it is also essential to understand that words within documents must be interpreted within the context of the document as a whole. To accomplish this, it is essential to understand the relationships between words. This understanding is fundamental to identifying structures and interpreting results.

1.3 Segmentation of the text chain into words

Text preprocessing is necessary to convert unorganized text into an organized and structured format, allowing machine learning methods to extract knowledge from textual data. Documents can be analyzed globally as a whole unit of analysis or divided into paragraphs, which are considered structured groups of sentences that maintain coherence on a specific subtopic.

In Natural Language Processing (NLP), tokenization is the process of dividing a text into lexical units, or tokens. Tokenization is a process that can be applied to separate sentences, words, characters, or subwords. Each result of tokenization is defined as a token. Given that in most cases tokens are words, the term *word* is used here in the broadest sense (more on this below).

The complete set of distinct words in a corpus is referred to as the lexicon. The basis for the statistical analysis of a corpus is counting the occurrences of the different *words* (or *tokens*) in the textual chain.

To carry out this process, it is highly recommended to eliminate any noise that the corpus may contain, as well as to normalize and standardize it. This phase is known as preprocessing.

1.3.1 Preprocessing the corpus

To mention only the simplest rules, it is necessary to:

- Perform a careful spell-checking using a powerful automatic tool that takes grammar rules into account.
- Indicate the characters that are considered as word separators, also called delimiters. Usually, these are the standard punctuation marks, and all other characters are treated as letters.
- Eliminate prescriptive notations such as capital letters at the beginning of a sentence or ambiguous abbreviations. However, it will be advisable to retain capitalization for proper nouns in order to distinguish them from homographic common nouns. Named Entity Recognition (NER) can identify words that correspond to people, organizations, and locations. This text normalization process can be completed by replacing non alphanumeric words (HTML tags, emoticons, hashtags...) with text, as well as removing URLs. Numbers can be replaced with their equivalent words or removed.
- Ensure that each character in the text has a unique status; for example, the period indicating the end of a sentence should be distinguished from the periods used in some abbreviations, such as U.N.
- Handling contractions (manually or automatically) in several languages by expanding them to their full forms.

1.3.2 Word, lemma, and syntactic category

A more precise definition of the term *word* is now necessary since the adopted counting rule must be stable, comprehensible, and reproducible. Furthermore, it must not vary during the preprocessing of a corpus or from one corpus to another and be easily understood by all users. Thus, it must be based on common knowledge. Some of these points are developed here.

Lexicographers have no other definition of the word than its typographical realization, that is, a sequence of letters delimited on the left and right by a delimiter (space or a punctuation mark). This definition also corresponds to either a *word-form* or a *graphic form*. It is equivalent to the term *token* used in NLP and machine learning. Converting a series of characters into a series of words (tokens) is called *tokenization*.

It is also common to identify the corresponding dictionary entry, or *lemma*, to which each word-form belongs. Identifying the lemmas requires:

- Grouping the different typographical representations of the same lemma (e.g., *want* and *wanted* are different realizations of the verb *to want* and thus both correspond to the lemma *to want*).
- Separating forms constructed by agglutinative procedures (e.g., separate *don't* into *do not* in English or *volverse* into *volver se* in Spanish).
- Distinguishing homograph forms (e.g., *content* can be a realization of the adjective *content* in *I'm content to have finished the book* or of the noun *content* in *the content of the book*; see Figure 1.1 and script **chap1.a**).

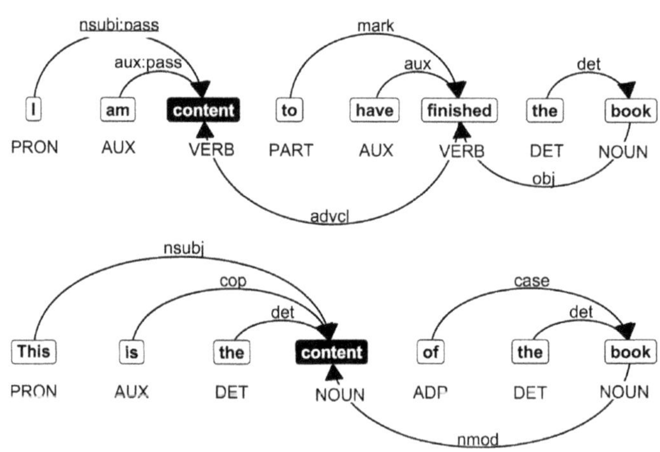

Figure 1.1: Dependency parser with **udpipe** and **textplot** R packages.

In most languages, lemmatizing vocabulary mainly requires conversion of the:

- Inflected forms of verbs to their infinitive form.
- Nouns to their singular form.
- Adjectives to their masculine singular form.

1.3.2.1 Lemmatization contribution

The lemmatization of the corpus before its analysis yields interesting results:

- It reduces the variability between documents. For example, when processing a corpus of free responses about attitudes toward reading, two children declaring, first, *I like to read a book* and, second, *I am someone who reads books*, they only share the *I* word-form, while they share the lemmas *read* and *book*.
- It associates each word-form or lemma with its syntactic category, which constitutes a fundamental contribution at different moments of processing. Thus, word-forms or lemmas can be selected according to their category, or the category can be used as illustrative information, i.e., as useful information in the interpretation phase.

Automating the lemmatization process requires a good morphosyntactic analyzer; even when using such a tool, ambiguities may still remain that can only be resolved manually.

For a given analysis, the user must select the type of lexical unit to be used. However, it is easy to repeat the same analysis starting from a different choice of lexical unit, because the statistical methods can be used without needing to make any special adjustments. This makes it possible to look at the same data from different angles.

In the following, *word* is used as a generic term, regardless of the lexical unit chosen by the user.

1.3.2.2 Information issued from a morphosyntactic analyzer

In general, a morphosyntactic analyzer produces a database containing, for each occurrence, the tag (grammatical category) and the lemma corresponding to it.

In this book we will use the **udpipe** package in **R** (R Core Team, 2025), which allows us to do tokenization, parts of speech tagging, lemmatization, and dependency parsing in different languages. Other packages in **R** are **tidytext** and **openNLP**. Part-of-speech tagging assigns each word to a grammatical category (tag), such as nouns, verbs, adjectives, and others, based on its meaning and context.

Here is an example corresponding to the document formed by the very short text in French *c'est de faire ce que l'on veut*. In this case, the morphosyntactic analyzer creates the structure shown in Table 1.1 where upos (tag) is the "Universal Parts Of Speech tag" (see script **chap1.a**).

Table 1.1: Overview of the data structure produced by the morphosyntactic analyzer.

id	token	lemma	upos(tag)	id	token	lemma	upos(tag)
1	c	ce	PRON	6	que	que	PRON
2	est	être	VERB	7	l	le	DET
3	de	de	ADP	8	on	on	PRON
4	faire	faire	VERB	9	veut	vouloir	VERB
5	ce	ce	PRON				

This data structure can be completed by adding two additional columns, the first containing the identification of the document number and the second the number of the paragraph to which the word belongs. This lemmatization process results in the creation of very large files with as many lines as occurrences, making direct analysis impossible and requiring this information to be synthesized.

As we will see later, this set of data can be utilized in a variety of ways. For instance, it can be employed to obtain statistics on the frequency of different grammatical categories or to select words according to their category.

It should be noted that all lemmatizers contain a certain percentage of errors, because, as Lebart & Salem (1998) commented, *the study of lemmatization problems shows that there can be no method which is both reliable and fully automatable for reducing to a lemma each of the units resulting from the segmentation of a corpus into word-forms*. Note that the grammatical category can also be used in a *supplementary role*, complementing the analysis based on either word-forms (tokens) or word-lemmas. As we will see in the following chapters, the factorial and clustering methods employed aim to reveal the underlying structure of the data. These methods are based on subsets of words and/or contextual variables and/or other information, collectively referred to as active data. Once the structure is determined, the remaining data can be used to illustrate it. This *supplementary* data is useful for interpreting the results.

1.3.3 Word selection

In textual data analysis, active words are usually chosen in advance for various reasons. For instance, in information retrieval, index optimization involves selecting a list of relevant words to maximize efficiency while taking the field and context into account. This selection may include single words, whether tokens or lemmas, or multiple words that meet certain criteria.

Next, we will comment on the frequency threshold and the grammatical or content roles that words play among the used criteria.

1.3.3.1 Frequency threshold

From a statistical point of view, comparisons between texts are only meaningful if they are based on words that are repeated with a certain threshold of frequency. In the case of bi- or multidimensional studies, it is usually advisable to establish a minimum frequency threshold for a word to be taken into account, as well as a minimum threshold of documents in which this word must appear. This is done to prevent a situation in which a small number of documents containing a word many times obscure the text's structure. Another option is to treat infrequent terms as supplementary elements.

1.3.3.2 Grammatical versus lexical words

According to Oxford reference, words whose primary function is to indicate grammatical relationships are referred to as *grammatical words*. This category includes articles, pronouns, prepositions, and conjunctions. Lexical or content words, on the other hand, serve a referential function and include nouns, verbs, and adjectives. However, the line is not always clear. For example, adverbs can be classified under either category. Depending on the objectives of the study and the type of corpus, we can choose to retain only words in one or the other of these two categories. As an alternative option, we can conduct successive analyses that include all words, followed by one category and then the other. Grammatical words are also referred to as function words or stopwords. The latter term is more pragmatic: stopwords are simply words that are not considered active during analysis. Selection criteria vary greatly depending on the chosen viewpoint. Grammatical words are usually the most common words in a language. However, they do not provide substantial information that would contribute to a more comprehensive text analysis. These words can be identified using TF-IDF (see Section 1.4.2.3), dictionaries, or by recognizing their grammatical category.

Removing these words ensures that only the most significant terms are selected. For some analyses, this means the matrices are smaller, resulting in lower computational costs.

1.3.4 Multiple words

In general, some sequential characteristics of text data can be captured by higher-order linguistic models, such as bigram or n-gram models.

1.3.4.1 Compound words

The process of segmenting text into smaller units can lead to controversial results when implemented in practice. Some compound words are lemmas in their own right, and are thus listed in dictionaries, which distinguishes them from syntagmas. Thus, in English, *purchasing power* is a dictionary entry (see English Robert and Collins French-English dictionary). But the French translation of this word, *pouvoir d'achat*, is not a separate entry in the same dictionary (French-English part) and is only mentioned as a compound word within the entry *pouvoir*. *Roadside* is a single word in English, but *bord de la route* is not even considered a compound word in French. It is what linguists refer to as a *fixed phrase*. It will be broken down into four words unless we are able to use an exhaustive list of such phrases.

In French, *s'envoler* (*fly away*) consists of two words, while *s'en aller* (*go away*) is three words since the same prefix *en* is attached in the first case but separated in the second. This is due to the variability in the French spelling standard, which admittedly contains some conflicting decisions.

In practice, a reasonable rule would be to follow usage, at least initially. Although some word-forms/word-lemmas may change their meaning depending on context (e.g., *peace in the world* versus *peace of mind*), our choice is to preserve either the word-form or the word-lemma as a single entity if the latter is identified as such by the morphosyntactic analyzer in use. We must accept that we have to get by with imperfect rules for the segmentation of the corpus into lexical units. For the types of problems we usually deal with, which generally involve non-linguistic issues, pragmatic solutions are entirely appropriate.

There are three types of compound words:

- The closed forms, where the words are a fusion, such as *notebook* and *basketball*, are not a problem in the analysis.
- The hyphenated forms, such as *mother-in-law*, present a choice, depending on whether the hyphen is considered a separator (three different words) or not (a single word).
- The open forms, such as *post office* and *middle class*, are repeated segments. In this case, they are bigrams. NER can be considered a special case of open forms.

1.3.4.2 Repeated segment

In this sense, in line with the work of Salem (1987), the detection of repeated segments makes it possible to locate compound words in the broadest sense. The definition of a repeated segment is simple: any sequence of words that are identical and not separated by a punctuation mark (i.e., a sequence delimiter). It should be noted that the segmentation of the corpus into word-forms or word-lemmas and the segmentation into repeated segments do not have the same status. The original text can be considered as a concatenation of the occurrences of words or lemmas, while the segmentation of the corpus into repeated segments produces overlapping lexical units (for example, the segments *I am* and *I am free* are two different overlapping segments; the respondent who said *I am free* used both segments). Our experience has led us to use repeated segments mainly as supplementary elements in the analyses based on word-forms or word-lemmas. These repeated segments can also be added to the co-occurrence table constructed from the terms.

The identification of repeated segments allows for the creation of a comprehensive dictionary of compound words present in the text, including how often they appear.

1.3.4.3 Quasi-segment

A quasi-segment is a sequence of words that are in the same sentence, repeated several times in the corpus but not necessarily contiguous. As with repeated segments, it is possible to obtain a dictionary of quasi-segments, as well as to extract only those in which a single word appears in order to understand the context of the word.

If the search space is a window of words rather than a sentence, the results will be sensitive to the size of the window for both repeated segments and quasi-segments.

1.4 Distributional theory of language

As Benzécri (1992) noted, the co-occurrences of words give meaning to the text.

This is consistent with Harris's distributional theory, which states that *the distribution of an element will be understood as the sum of all its environments* (Harris, 1954, p. 146). Treating documents as lexical distributions enables us to compare them based on the relationships between their words. In the following sections, we will explore how to consider the context of words. We will start with the most basic approach of focusing on concordances of privileged words and then progress to the most elaborate approach of starting from a co-occurrence table and analyzing it using specific methods.

1.4.1 Concordances of privileged units

A concordance of a selected word (keyword) or repeated segment is a list that shows the immediate context and its locations in the text (document identification and line or sentence number). These tools facilitate the discovery of how the same keyword is used in different situations, allowing us to understand its meaning by relating it to the context in which it appears (see Table 1.2 for an example).

When the number of instances of a word or term is high, it can be challenging to identify the different contexts in which it is used. In such cases, the use of graphics, such as WordTree, facilitates interpretation by providing a visual representation of the data (see Figure 1.3).

1.4.2 From corpus to crossed-tabs

1.4.2.1 Table of co-occurrences

Adopting a distributional approach to a corpus results in its representation as a table of word co-occurrences. By examining how words appear together in a document or in close proximity, we can gain insights into their usage and connotations. It is a useful method to identify and quantify the semantic associations between words. The proximity can be determined by the size of the context, establishing the number of words to be analyzed before and after the target word. There are three main types of non-exclusive co-occurrences, which are usually provided in lists with the corresponding frequencies:

- Words co-occurring within the same document or paragraph.
- Words co-occurring within a maximum distance (skipgram) of a target word, which defines that word's neighborhood.
- Immediately adjacent words (a special case where the skipgram is 0), which can form repeated segments.

The results are typically presented in tabular form, in which each co-occurrence is accompanied by its respective frequency. This symmetric table has both rows and columns labeled with the different words. In the corresponding cells, we have the frequency with which the two words co-occur within the chosen context.

There are many ways to analyze this co-occurrence table. We will discuss two methods: the GloVe vectorization method (see Section 1.5) and correspondence analysis (CA) (see Chapter 2).

1.4.2.2 Document × Term table

Another way to identify word co-occurrences is to use cross-tabulation, which allows for a comparison of the overall word distribution across different documents. The frequency of each word in each document is contained in the corresponding cells of the table crossing documents and words, called either the *document term matrix* (DTM) or the *lexical table* (LT). This corresponds to the typical table analyzed by CA. In Chapter 2, we discuss how latent semantic analysis (LSA) and CA allow us to identify patterns and relationships between words and documents by applying them to this table.

1.4.2.3 TF-IDF table

The DTM table can be transformed to assign importance to words based on their ability to differentiate documents. This is how a TF-IDF table is constructed.

The raw count definition of *Term Frequency* (TF) is the number of times the term or word j appears in a document i: $n(j, i)$. The standard definition of TF is the raw count $n(j, i)$ divided by the total number of words in the document i:

$$\text{TF}(j, i) = n(j, i)/I_i.$$

The logarithmic scaling formula of TF is a variation intended to mitigate the impact of high frequencies:

$$\text{TF}(j, i) = log(1 + n(j, i)).$$

The term *Inverse Document Frequency* (IDF) is the logarithm of the quotient between the number of documents in the corpus I and the number of documents in which the term j appears I_j:

$$\text{IDF}(j) = log(I/I_j).$$

If the term j is present only in one document, $\text{IDF}(j) = log(I/1) = log(I)$. If the term j is present in all documents, $\text{IDF}(j) = log(1) = 0$.

The IDF value increases as the number of documents containing the term decreases. This approach ensures that words that are unique or present in a small percentage of documents receive higher importance IDF values than words common across all documents, such as articles, conjunctions, and prepositions.

The measure known as *Term Frequency – Inverse Document Frequency*:

$$\text{TF} - \text{IDF}(j, i) = \text{TF}(j, i) \cdot \text{IDF}(j).$$

Higher values of TF-IDF are related to high frequency terms in a document and rare in the corpus.

Most frequencies in texts come from common words. These words tend to provide little information except for stylistic analysis, and they slow down the analysis process. It may be advisable to remove these words by establishing a maximum frequency, or to use a more precise methodology, which is discussed in the following section. This table is used in several methods. Among them, we can cite LSA (see Chapter 2).

1.5 Text vectorization

In the field of NLP, various methods have recently emerged that enable efficient data storage using different data structures. The structure closest to the original text contains each token in a row (see Table 1.1), including punctuation characters. This large file can often be transformed into other structures that allow statistical methodologies to be applied.

Vectors in textual data are numerical representations of words, phrases, or documents (in rows) that allow computers to perform statistical operations on them. *Vectorization* is the process of transforming any data type as vectors.

This process must be carried out according to the objectives. It can be classified as follows:

- One-hot encoding. This method generates a unique vector for each word in the vocabulary. The vector's length is fixed and equal to the total size of the vocabulary. It contains a 1 at the index representing that word and 0 in all other positions. This method is typically used as an input for neural networks, not for building a DTM.

- The Bag of Words (BoW) is a model that represents a document as an unordered set of its words, disregarding grammar and word order. BoW is the basis for DTM. A *count vectorizer* is a modification of the BoW model, where the vector's elements are not binary; instead, each number represents the frequency with which a corresponding word is present in a document. The result of representing all documents as row vectors is a DTM, which has as many columns as the combined vocabulary of all documents. This creates a high-dimensional sparse matrix. One limitation of the DTM is the loss of word order. Consequently, meaning and context are often considered uncapturable. However, it is possible to recover part of the relationship between words by reducing the dimensionality, which is important for understanding the text (see Chapter 2).

- Weighted DTW. Although TF-IDF is often considered a vectorization method, it is actually a type of matrix weighting applied to a DTM matrix (see Section 1.4.2.3). Chapter 2 covers other DTM weighting methods. In practice, data structures and algorithms for sparse matrices can be represented using simple triplet representations. Each row in the matrix corresponds to a non-zero frequency value. The first element is the row index, the second is the column index, and the third is the value of the matrix cell (see Hornik et al. (2009) for an example using the **slam** package).

- Word embedding. This term encompasses a variety of models that share several types of information-gathering, storage, and methodologies. We distinguish between static and dynamic (or contextual) embeddings. Static embeddings build a single row vector for each word in the vocabulary at a specific time, such as during the learning phase. Once a static embedding vector representing a word is computed, it remains unchanged, regardless of the context in which it is used. This can be a problem when words can have different meanings than the context of training data.

Some of these algorithms are implemented in **R** packages such as **word2vec**, **GloVe**, **textrank**, and **fastText**.

- The **word2vec** package offers two main types of static embedding algorithms: Continuous Bag-of-Words (CBoW) and Continuous Skip-Gram (SG) (Mikolov et al., 2013). They use a deep learning-based approach where words are represented in rows and the columns correspond to the dimensions (or vector size) obtained from the trained dataset. The cells contain the real number value of the word for the respective dimension, which is determined by the specific algorithm applied.

Although a larger vector size can capture more semantic information, it requires more memory. The number of dimensions is usually much smaller than the original vocabulary size—usually no more than 300—and is typically lower in specific environments than in more general contexts. Clearly, the quality of the retrieved information depends on the number of chosen dimensions. However, a large number is not always the best solution because it can retrieve random information that does not help one to understand or predict, especially with small data as open questions.

Therefore, little storage capacity is required: as many rows as there are words in the vocabulary, and as many dimensions as there are columns. For this reason, they are said to obtain dense representations, compared to other sparse representations such as DTM.

They are also referred to as *context window methods* because they utilize a window of predefined size to define the context words surrounding a target word. The typical window size is five for CBoW and ten for SG.

CBoW is a neural network-based algorithm with the objective of predicting a target or center word (row) using its surrounding context words (defined by the window size) as input. Previous methods of representing data lost the proximity relationship between words in a document.

On the contrary, the SG objective is to predict the surrounding context words around a target word. The simplest type of n-grams is the bigram, which represents a combination of two consecutive words in a text. For example, in the negation *not bigger*, the token *not* is separated from *bigger* in a BoW approach. However, the bigram is able to retain this local sequential information. This kind of proximity information, which is entirely lost in sparse representations like BoW, is crucial for capturing semantic meaning. When higher-order n-grams are applied, the number of possible word combinations grows exponentially. This results in significant sparsity and high dimensionality, requiring massive memory, processing, and storage space. On the other hand, it does not make sense to rely on the resulting combinations when their frequency is very low, especially when applied to small datasets. In general terms, both are computationally efficient, although CBoW is recommended for high-frequency target words and SG is recommended for infrequent or rare ones. The training neural network can be applied using a hierarchical softmax (better for infrequent words), or using negative sampling (recommended with high-frequency words, small vectors and large datasets). These algorithms can typically be used in combination with the selection of words that have a minimum frequency, as well as with the establishment of stopwords.

- **fastText R** package. While Word2Vec takes each word as a single unit (row), **fastText** (Mouselimis, 2021) considers each target word as the sum of all its character n-gram vectors (subword units). For example, the word *book* in Figure 1.1 can be represented by a Bag of Characters n-grams (subword units) and n=3, using < symbol as the left of the word and > as the end of the word: <bo, boo, ook, ok> and <book>. This method is more computationally intensive because it must store vectors for both words and their n-grams. However, it offers the advantage of being able to generate a representation for out-of-vocabulary (OOV) words, for example, if the word *booking* were encountered.

- **GloVe** (Global Vectors) model. Unlike Word2Vec models, which rely solely on local context using windows, the GloVe model analyzes the entire corpus directly. GloVe is implemented in **R** using the **text2vec** package (Selivanov et al., 2016). It breaks the text into tokens, applying the criteria set out in Section 1.3.3, and builds a co-occurrence matrix words × words. The value of each cell indicates how many times the word i appears in the same context as the word j. This value is typically weighted by the inverse of the distance between the two analyzed words, giving more weight to words that are closer together. The context is generally defined using a sliding window. We can also consider a third word k and calculate if it is more closely related to i or to j.

  This co-occurrence table is highly sparse, so it is reduced in size using the **Matrix** package. In order to reduce the number of dimensions, **text2vec** employs a stochastic gradient descent algorithm. Compound terms (repeated segments) within the window are detected and can be incorporated as compound words in the co-occurrence matrix. This methodology, based on the co-occurrence matrix, also allows for the identification of the closest word vectors for a given word.

Among the dynamic algorithms that are beyond the scope of this book, we highlight BERT (Bidirectional Encoder Representations from Transformers developed by Google) and ELMo (Embeddings from Language Models).

In some cases, the vectorization process does not allow for or hinders individual analysis of documents. Starting from the above representations of words (such as Word2Vec or GloVe) as a starting point, the difference or similarity between two documents can be measured by aggregating the word vectors (taking the mean of all word vectors in the document or weighting using TF-IDF) and then comparing the resulting document vectors. Another possibility is to use the **text2vec R** package, which allows for preprocessing text data, creating a DTM, and implementing advanced text vectorization techniques.

Text vectorization is a fundamental part of NLP, transforming a corpus into a set of numerical representations. This way, algorithms can process and analyze the data. Statistical analysis can be applied to text data to synthesize information and discover complex structures, both from the perspective of words and documents.

1.6 Extraction of characteristic elements

A keyword can be defined as a word or phrase that represents the primary topic or concept in a text, providing a summary of its content. This text summarization feature enables users to search for and retrieve specific information. The most basic form of identification is based on frequency-based methods, which is why it is often closely related to co-occurrences.

1.6.1 Methods for extracting keywords

KWIC (Keyword in Context) is a text analysis technique that involves from a simple display of concordances until to identify keywords, showing when two or more words appear together (see example in Figure 1.8). For this reason, different KWIC NLP techniques can be considered closer to concordances than to co-occurrences. For instance, the **R** package **udpipe** includes a function called keywords_collocation that compares the probability of two terms appearing together to their statistical independence, and **quanteda** has a function named kwic to do concordances and co-occurrences.

KWIC also provides graphical networks to help understand the relationships between words and numerical statistics to measure the similarity between two words. We will use three techniques under the name KWIC from the **udpipe** package: RAKE, PMI, and SNP.

RAKE (Rapid Automatic Keyword Extraction) (Rose et al., 2010), splits the text using punctuation marks, irrelevant words (stopwords) or grammatical categories as delimiters, extracting candidate keywords from this continuous sequence of words removing irrelevant words. A logical vector named *relevant* can be built to define the analyzed words. Each word has a RAKE score, indicating how often it appears with other words related with the frequency of the keyword (udpipe::keywords_rake function). It is also often possible to define relevant keywords to narrow down the search.

PMI (Pointwise Mutual Information) has the objective of extracting grammatically or semantically fixed expressions (collocations), enabling the selection of the grammatical type of words to analyze. It is implemented in the udpipe::keywords_collocation function. It provides three statistics to obtain the results: PMI, MD (Mutual Dependency), and LFMD (LogFrequency biased Mutual Dependency):

$$\text{PMI}(w_1, w_2) = \log_2\left(\frac{P(w_1 \cap w_2)}{P(w_1)P(w_2)}\right).$$

$$\text{MD}(w_1, w_2) = \log_2 \left(\frac{P(w_1 \cap w_2)^2}{P(w_1)P(w_2)} \right).$$

$$\text{LFMD}(w_1, w_2) = MD + \log_2 \left(P(w_1 \cap w_2) \right).$$

SNP (Simple Noun Phrases) is the process of extracting noun phrases from text. This can be achieved by defining a combination of several types of sequences of Parts of Speech (POS) tags in the mboxudpipe::as_phrasemachine function (Handler et al., 2016).

1.6.2 Characteristic words and documents (LexChar)

This section describes how to characterize the different parts of the corpus using characteristic words and documents. The parts of the corpus can be determined using various methodologies. For example, the corpus of *Spanish investiture speeches* is classified into categories based on the type of speech and the politician who delivered it.

The *Health* corpus consists of responses to an open-ended question and can be grouped according to the categories of a closed-ended question, such as gender. These responses can also be grouped according to clusters obtained using a classification method. If we are dealing with a long corpus composed of a single text—such as the *Closing argument* example—we can divide it into parts according to chronology. The goal is to identify the characteristic words of each part. Instead of identifying them separately, we want to identify them by comparing the parts. The documents of a DTM can be characterized in terms of the words that are significantly over- or underrepresented in each of them. These words are considered as keywords within the documents. This makes sense for large documents but not for short answers to open-ended questions, or other types of numerous short documents. In this case, the answers must be grouped by the categories of a qualitative variable or by the groups resulting from a clustering analysis.

Starting from the DTM, an aggregated lexical table (ALT) is created by combining answers in the same category or cluster into one row. This is easily done by using the function LexChar of the package **Xplortex** (Bécue-Bertaut et al., 2024). In order to characterize each category or cluster, it is helpful to identify the words that are significantly over- or underrepresented in them. Note that eliminating hapax legomena and infrequent words ensures that the characterization is not influenced by rare words that do not generally correspond to the analyzed topic's context. See Lebart et al. (1998, chap. 6) and Bécue-Bertaut (2019, chap. 5) for a detailed description of the calculations and statistical tests used to determine the significance of the obtained values. For an example of the results and their presentation, see Section 1.8.8.

To identify the most relevant documents based on keywords, the **textRank** package uses the textrank_sentences function to compute the Jaccard distance between keywords and sentences applying the Google's pagerank algorithm (Wijffels (2017), Mihalcea and Tarau (2004)).

1.7 Examples

We plan to demonstrate these methods by applying them to real-world examples. However, we will not conduct an in-depth study of how to interpret the results.

Here, we develop a brief overview of the types of corpora used throughout the book to give a glimpse of the variety of corpora that can be analyzed using a multidimensional statistical approach. However, we are far from having compiled a list of all types of corpora that can be profitably treated with these methods.

1.7.1 Example Health: non-expert definitions of health

An international survey was conducted in 1989-1990 to understand the attitudes and opinions of non-experts regarding health in order to improve public health policy. Here, we present some of the data collected from a medium-sized coastal town in Valencia, Spain. A sample of 513 people over 14 years old was observed. The first question, *What does health mean to you?*, required free and spontaneous responses. We have the open-ended responses to this question, as well as the closed-ended socioeconomic questions. The aim is to identify the different definitions of health and how they vary according to socio-economic characteristics, in particular age, gender, and socioeconomic level. The *Health* corpus consists of all the free responses and their associated closed responses.

1.7.2 Example Children and Reading: A survey about children's reading opinions, attitudes, and practices

A questionnaire concerning reading habits was answered by 895 fifth-grade children. This questionnaire included two open-ended questions: *Para mí leer es... (For me, to read means)* and *Creo que leer es importante porque... (I believe that reading is important because).* It also included closed questions about opinions, attitudes, and practices regarding reading as well as socioeconomic characteristics and academic performance.

1.7.3 Example Train: satisfaction survey about night travels by train

To gain a comprehensive understanding of customer satisfaction with night rail services, a Spanish railway company collected passenger feedback.

A random sample of international trips was selected, and all of the travelers were asked to rate 14 specific aspects on a scale from 0 (very poor) to 10 (excellent). These aspects served as contextual variables. Then, via an open-ended question, respondents were asked to identify areas for improvement. Respondents answered in either Spanish or English.

1.7.4 Example Closing Argument: prosecutor's closing argument in a murder trial

A forensic speech aims to persuade the audience (whether judges, the court, or the jury) of the defendant's guilt or innocence regarding a crime, all within the confines of the law. In a courtroom, rhetoric primarily appeals to logical reasoning to support and connect all arguments effectively, giving the speech the form of a demonstrative speech. Here, we will examine the closing argument delivered by a Barcelona prosecutor who accused two accomplices of murdering a prostitute they employed.

1.7.5 Example Wines

This example falls within the field of sensometrics, which is the statistical analysis of sensory product evaluations. In this example, the corpus consists of verbal data collected during a test session organized with wine experts according to a precise protocol.

1.7.6 Example Spanish investiture speeches

The database contains investiture speeches delivered by candidates for Spanish prime minister since the restoration of democracy. This study focuses on eighteen political speeches, each with approximately 10,500 occurrences. The texts of the speeches were taken from the published versions of the *Diario de Sesiones del Congreso de los Diputados* (Journal of Sessions of the Congress of Deputies), published by the Spanish Parliament. The corpus is identified with the name *SpanishDisc18*.

1.8 Lexicometric results

The following sections present the results obtained in the *Health* example (Section 1.7.1), which asks *What does health mean to you?*

The analysis begins with the importation of the Health.0 file, which contains the original responses to the open-ended question that have been translated into English (script **chap1.b1**). The text uses lowercase letters. Capital letters have only been kept for proper nouns and the pronoun *I*.

Abbreviations are checked and replaced (script **chap1.b2**). In this corpus only the abbreviation *i.e.* was changed to *ie*. Contractions are replaced by their long form using replace_contraction function of **qdap** package (Rinker, 2023) (script **chap1.b3**). Symbols were checked and replaced by text (script **chap1.b4**). In this preprocessing phase, only the symbol % was changed to *percent*.

The **udpipe** package was used to perform POS tagging and lemmatization on the text. The result was saved under the Health name (script **chap1.c**).

1.8.1 Corpus size and lexicon size

The TextData function of the **Xplortext** package reads the preprocessed file and builds the DTM (script **chap1.d**). The DTM has as many rows as there are nonempty documents (I=394) and as many columns as there are different words (J=vocabulary size=924). At the intersection of row i and column j, the (i, j)th element indicates how many times the document i uses the word j. In total, there are 10,301 occurrences in the text, with a mean length of 26.14 words for nonempty responses.

In order to remove stopwords, **Xplortext** retrieves the list of 174 English stopwords from the **tm** package. The pronoun *I* that remained in uppercase was also removed using the user-defined stopwords list. The 11 most frequent words are stopwords (*to*, *is*, *and*, *not*, *the*, *have*, *a*, *you*, *be*, *that* and *of*). They account for 35.4% of the total occurrences. After removing stopwords and numbers, the corpus size decreased to 4,274, the vocabulary size to 811 and the mean length of the responses to 10.88.

It is also possible to remove words that are used less than a certain threshold (Fmin) or that are not used in at least a minimum number of documents (Dmin), or that are used too frequently (Fmax).

1.8.2 Index of words and repeated segments

The results in this section correspond to the responses to the open-ended question, after preprocessing and removing stopwords but before lemmatization.

Figure 1.2 lists the 25 most frequent words. It reveals positive aspects such as *health*, *well*, and *good*, as well as negative aspects such as *pain*, *ill*, and *illness*, related to *physical/physically* and *mental* health.

Repeated segments provide part of the context of words, helping us to understand them. Note that these segments are an integral part of the corpus and therefore include stopwords. Some examples of segments and their frequency are *to be (212)*, *not to (110)*, *to have (101)*, *to feel (93)*, *not have (91)*, and *it is (90)*, corresponding to generic expressions of verbs commonly found in a traditional corpus. However, certain repeated segments can retrieve relevant information from the provided responses related to the context as: *to feel (93)*, *have any (80)*, *feel good (75)*, *to feel good (63)*, *be well (74)*, *to be*

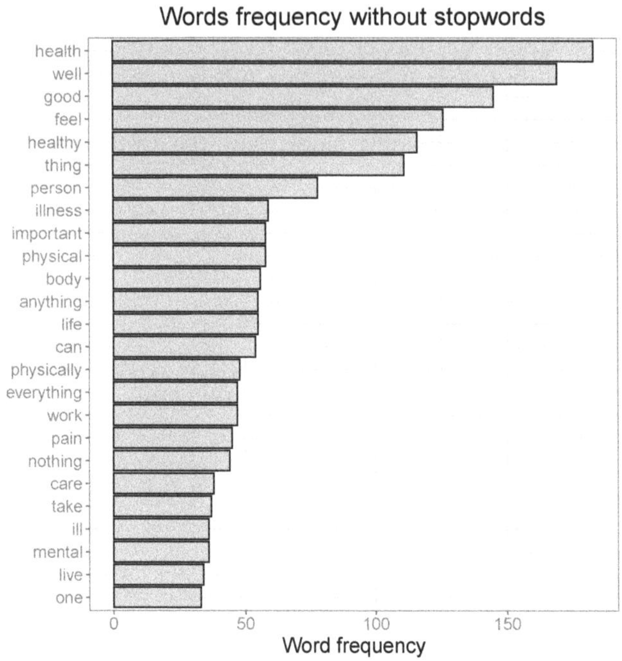

Figure 1.2: Most frequent words after removing stopwords.

well (71), *do not (70)*, *health is (61)*, *not to have any (59)*, *any illness (45)*, *have any illness (43)*, and *not have any illness (42)*.

1.8.3 Concordances

Concordances can help you better understand the context in which a word is used and accurately interpret its meaning. For example, the lemma *drink* appears as the words *drink (20)*, *drinking (3)*, and *drinks (3)*. Knowing the context is essential to fully comprehend this lemma in relation to *health*. To accomplish this, the text containing *drink** is selected so that it includes all variations starting with this word stem. This allows the immediate context to be shown. In this case, with a window size of five, the text to which the *drink** belongs along with its position is displayed (Table 1.2).

As can be seen, most of the occurrences of *drink** are related to *alcohol*, *tobacco*, *drugs*, and *exercise*.

As shown in the interactive Figure 1.3, created using quanteda::kwic (Benoit et al., 2018) and googleVis::gvisWordTree (Gesmann and de Castillo, 2011), the term *drink* (occurring 20 times) is most closely associated with the term *alcohol* (16 occurrences). The word tree can represent the context before or after the word *drink*, either independently or simultaneously.

Table 1.2: Concordances for *drink∗* words.

Text and position	Before	Word	After
[t3,13]	not to smoke, not to	*drink*	alcohol, to eat properly
[t8,12]	a person who not to	*drink*	alcoholic beverages, not to smoke.
[t12,37]	who does not smoke or	*drink*	alcohol and grooms himself well.
[t18,49]	, advice, tobacco, drugs,	*drink*	alcohol, which are the source
[t23,7]	not to smoke, do not	*drink*	alcohol, if you have time
[t59,19]	well and does not smoke or	*drink*	alcohol.
[t75,10]	is healthy, who does not	*drink*	does not take drugs
[t232, 8]	a healthy life, not to	*drink*	alcohol and to do exercise and
[t245,9]	with meals and, neither	*drink*	alcohol, nor eat spicy things
[t254,142]	even the water can not	*drink*	, we have to buy it
[t302,8]	always follow a diet and	*drinking*	regime, to do exercise,
[t302,41]	thing, both in terms of	*drinking*	, eating, not to smoke
[t304,41]	foods harmful to health, or	*drink*	alcohol.
[t306,4]	is not to	*drink*	alcohol, not to smoke and
[t307,6]	it would be not to	*drink*	alcohol, not to smoke,
[t313,26]	quit smoking and not to	*drink*	alcohol and to do sport,
[t316,14]	hurts you, try not to	*drink*	alcohol, not to smoke,
[t437,59]	who does not smoke or	*drink*	alcohol, who do exercise every
[t442,12]	for himself, eat anything,	*drink*	anything and live a life like
[t448,12]	is harmful to smoke, to	*drink*	alcohol, joints and drugs,
[t459,7]	not to smoke, not to	*drink*	alcohol, do not take drugs
[t461,45]	over your body in food,	*drinks*	, sleep and so on.
[t473,35]	, because if we run,	*drink*	and carry a rhythm like that
[t497,20]	, is not to smoke, not,	*drinking*	.

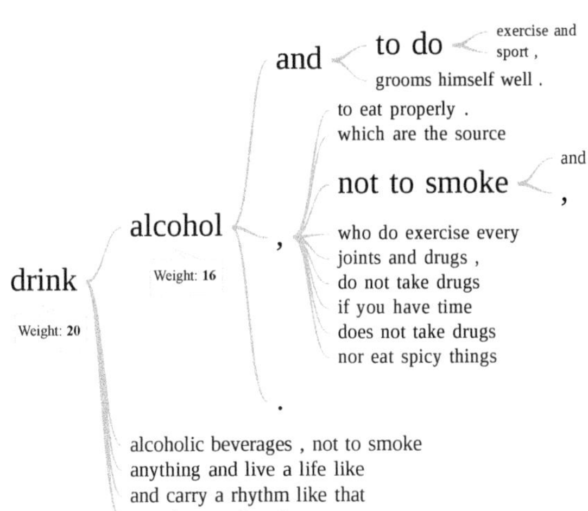

Figure 1.3: Context of *drink* word. WordTree.

1.8.4 Lemmatization

Lexicons may indicate the part of speech (e.g., noun, verb, adjective) that a word belongs to, helping users understand how the word is used grammatically. The analysis of the lemmatized text (script **chap1.e**) can be performed using the grammatical categories of the lemmas saved in the upos variable in the dataframe constructed using **udpipe** in script **chap1.c**. Assuming that df is the dataframe, the statistics of the lemmatized text can be calculated as: udpipe::txt_freq(df$upos). The second option is to work directly on the lemmatized text.

Once the text is lemmatized, it is possible, for example, to select POS information (Figure 1.4) providing a summary of the frequency of each grammatical category. It should be noted that the highest frequency of occurrence is observed in punctuation marks, which will be removed during the statistical analysis. This is followed by nouns, verbs, pronouns, and adjectives.

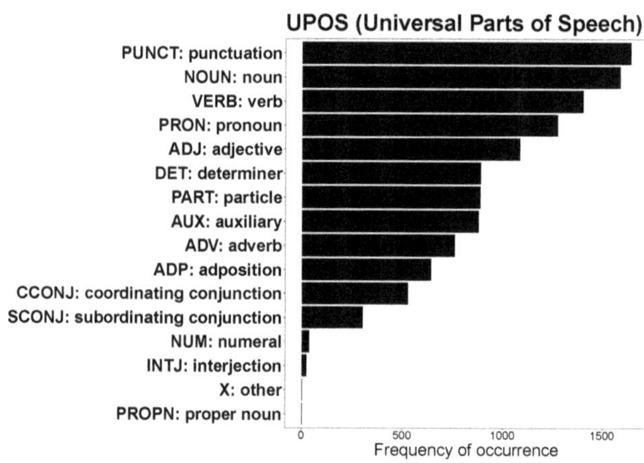

Figure 1.4: UPOS frequency (Universal Parts of Speech).

The syntactic category of verbs (lemma) can be compared with the unlemmatized text (token) and obtain frequency tables and graphs for each grammatical category (Figure 1.5).

1.8.5 Vectorization

As indicated in Section 1.5, count vectorizer is a modified BoW structure in which the number represents the frequency with which a word appears in a document. The BoW representation or the DTM matrix can be obtained directly from the text using the Xplortext::TextData function. This yields 394 documents, excluding empty ones, and 924 columns corresponding to the vocabulary before removing stopwords. The value in each cell is the frequency (script **chap1.f**).

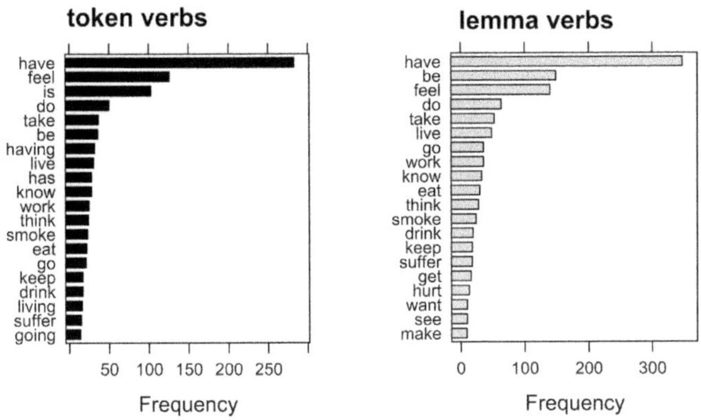

Figure 1.5: Most frequent token and lemma verbs.

The training data set was created by removing stopwords and setting a frequency threshold of five, resulting in a total of 144 words. The findings for a set of five words related to the term *drink*, derived from a model with fifteen dimensions, are summarized in Table 1.3. However, the dimensions of this table are abstract and cannot be interpreted directly. It should also be noted that the dimensions are not associated with an ordered importance related to the explained inertia.

Table 1.3: Four dimensions from CBoW.

Words	D1	D2	D3	D4
alcohol	0.32	1.64	−1.60	0.78
condition	−1.10	−0.82	1.19	−0.53
drink	−1.29	−0.90	−0.59	−1.62
health	0.46	0.72	0.13	−0.92
smoke	0.30	1.33	−0.50	0.76
sport	−0.67	0.94	0.02	−1.55

Representing the relationship between words in a 2D or 3D map requires reducing the 15 dimensions using a technique like t-SNE or UMAP (see Chapter 5 for a description of these plots). Script **chap1.f** includes several representations in 2D. For example, Figure 1.6 illustrates the relationships connecting one group of words (*exercise*, *body*, and *alcohol*) with another (*drink*, *smoke*, *sport*, and *tobacco*).

Skipgram model in the **word2vec** package provides a similar intermediate structure words × dimensions. It is important to note that ensuring reproducibility using the Word2Vec model starting from the same data is not always possible. Since this embedding process can be computationally expensive and to avoid the problem of no reproducibility, it is advisable to store this matrix for later processing.

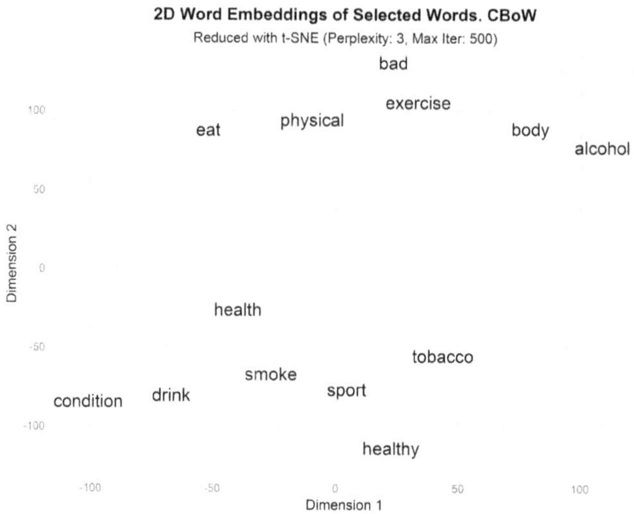

Figure 1.6: 2D-Word embeddings of selected words.

Another option is to create a data frame with one row for each term using **udpipe** (script **chap1.e**). In our case, that would be 11,947 rows. The order of words, including punctuation marks and stopwords, is maintained as they are written in the text. Each row contains the identifier of the document to which the word belongs, as well as the paragraph and sentence. In contrast to DTM, it is possible to quantify the distance between two words in a document by taking into account the number of words separating them. This approach enables enhanced precision of the analysis by segmenting documents into distinct components. It is also possible to perform analyses by selecting grammatical categories.

The package **text2vec** incorporates an implementation of the GloVe algorithm, which provides a matrix words × dimensions with a structure analogous to a CBoW. Additionally, it includes a sparse matrix of words × words.

1.8.6 Co-occurrences

Co-occurrence analysis provides insights into how words are used in proximity to one another within a document, sentence, or paragraph. The relationships between words can be summarized in a table and visualized in maps. In this section, the lemmatized text is used in order to analyze the grammatical categories in the form of co-occurrences.

To identify repeated two-word segments, we can use the udpipe::cooccurrence function to find co-occurrences within certain grammatical categories (script **chap1.f**). For instance, setting skipgram to zero will find adjacent nouns and adjectives (script **chap1.g**).

Table 1.4 lists the twelve most frequent repeated two-word segments and their frequency. Among them, the word *thing* appears four times and *health* three times.

Table 1.4: Repeated segments using only nouns and adjectives.

	Term1	Term2	Cooc		Term1	Term2	Cooc
1	important	thing	36	7	healthy	person	11
2	main	thing	28	8	mental	health	11
3	best	thing	21	9	good	health	8
4	healthy	life	12	10	little	bit	6
5	physical	fit	12	11	many	thing	6
6	good	nutrition	12	12	physical	health	5

In addition to providing contextual information about words, co-occurrence analysis is a flexible tool for identifying topics (Figure 1.7). Representing only nouns and adjectives (left), *health* is at the center of the map, getting words already mentioned above. There are also words such as *job*, *money* (without *health* there is no *job* and no *money*), *nutrition*, *able*, *ill*, and *illness*.

To the right of the Figure 1.7, the consideration of verbs alters the overall configuration. Verbs (in bold) are frequently placed in the center. Words appearing in both figures are italicized. Only a few new words appear: (*doctor*, *much*, *kind* and *year*). The verb *have* is in the center, because it is related to very different words. The proximity of verbs to other words provides information about their meaning. For instance, *good* is next to *feel* (coming from *feel good*), and *care* is next to *take* (coming from *take care*).

These co-occurring words help us understand documents where *health* appears (*healthy life*, *healthy person*, *health you*, *good health*, *physical health*, and *mental health*). Other co-occurring words that do not include the word *health* (such as *mental condition*, *physical fit*, *your body*, *eat properly*, *good nutrition*, *drink alcohol*, *some sport*, *nothing hurt*, and *take care*) also help us understand the content of these documents.

One way to consider only the strength of the relationship between words, while eliminating the influence of frequency, is to use correlations or cosines. With a minimum frequency set at five, the highest correlations between two words were found for *hygiene personal (0.89)*, *care take (0.81)*, *mental physical (0.75)*, *base hygiene (0.74)*, *base personal (0.68)*, *drink smoke (0.68)*, *affect day (0.65)*, *enough happiness (0.63)*, *child see (0.59)* and *feel good (0.58)*. The strongest correlations between *drink* and other words (including nouns, adjectives, and verbs) were found with *smoke* (0.68), *alcohol* (0.52), *drug* (0.46), *sport* (0.25) and *eat*(0.23).

As previously mentioned, the **text2vec** package uses a stochastic gradient descent algorithm that can calculate the closest word vectors to combinations of words. For example, it can find the vector closest to the sum of the drink and alcohol vectors (script **chap1.h**). The words with the highest cosine similarity

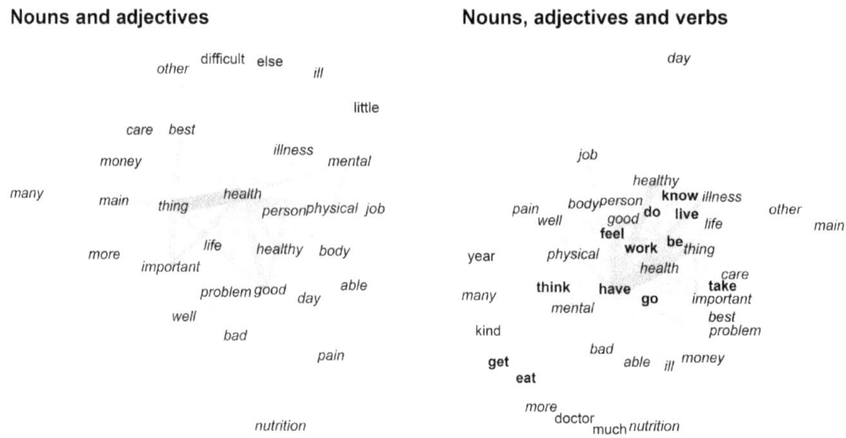

Figure 1.7: Co-occurrences Fruchterman-Reingold algorithm. Verbs in bold, shared words in italic.

to this combined vector (*drink + alcohol*) are *drink* (0.94) and *alcohol* (0.93), followed by *smoke* (0.61), *get* (0.41), *suffer* (0.39), and *always* (0.39).

The **text2vec** package also facilitates the extraction of co-occurrence structures using PMI and the addition of these repeated segments to the sparse co-occurrence matrix of words $\times$ words. It detects 70 repeated segments over a frequency of four, with the most significant using PMI: *something_very_important*, *little_bit*, *no_pain_at_all*, *or_anything_else*, *drink_alcohol*, *suffer_from*, *going_out*, *too_much* and *take_care_of_oneself*.

The script **chap1.f** provides an example using **vec2word** which utilizes embedding skip-grams methodology to identify significant relationships between *alcohol* and *drink* and the rest of words, finding related concepts as *smoke*, *sport*, *eat*, *exercise*, *do*, *does*, and *living*.

1.8.7 Global keywords

As explained in Section 1.6.1, **udpipe** provides algorithms for extracting keywords from a corpus. These algorithms identify frequent word associations that meet specific criteria across the entire corpus.

To illustrate this process, three different algorithms are used to extract keywords consisting of no more than two words. Only nouns and adjectives are considered (script **chap1.i**). Figure 1.8 shows the 20 most representative keywords from the RAKE, PMI, and SNP algorithms, respectively. Note that they characterize the entire corpus.

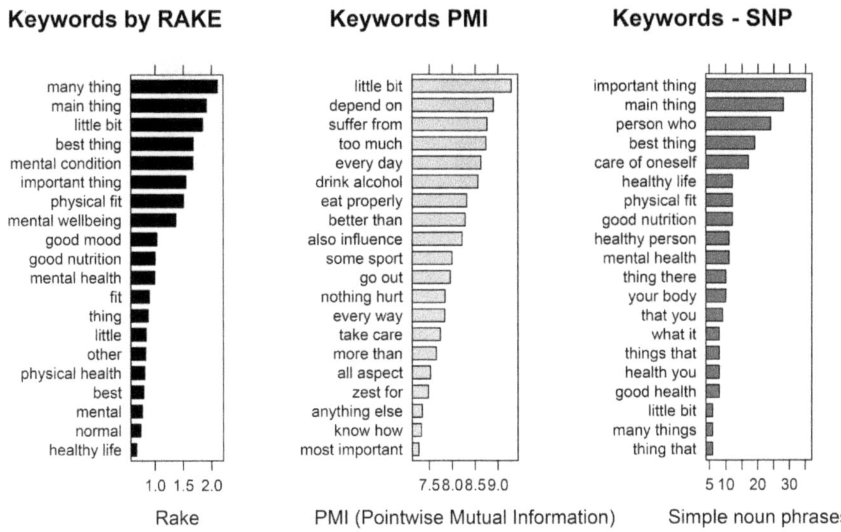

Figure 1.8: Keywords selected by **udpipe R** package.

RAKE uses the function keywords_rake. The word *thing(s)* is a component of several high-scoring keywords, including *many thing*, *main thing*, *best thing*, and *important thing*. *Good* is related to *mood* and *nutrition*, *health* with *mental* and *physical*.

The mboxkeywords_collocation function identified several noteworthy keywords using PMI, including *drink alcohol*, *suffer from*, *some sport*, *nothing hurts*, *take*, and *taking care*.

1.8.8 Keywords of corpus parts

The goal is to characterize the different parts of the corpus using keywords. We will use the gender variable to divide the health example into two parts. Using a minimum threshold of eight words (Fmin=8), an ALT is built with two rows (referred by man and woman labels) and 96 words-columns (script **chap1.j**). Figure 1.9 shows the keywords by gender, also called characteristic words, using the Xplortext::LexChar function.

Men's responses are related to *general* aspects, physical aspects (*fit*, *sport*, *exercise*, *not to smoke*) and *health* as a *way* of life and to *keep* well. Women indicate that *health* is the best *thing* in the *world*. They also overuse certain negative words (*hurts*, *pain*, *nothing*) and positive ones (*feel*, *feels*, *good*, *well*, *everything*). Additionally, women use the word *now* in a similar context as men use the word *keep*.

To understand the meanings of the over- or underrepresented words for a given category within a context, it is interesting to obtain the words that

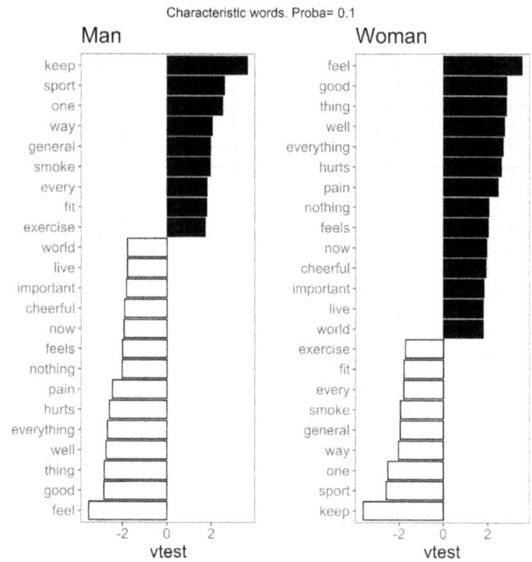

Figure 1.9: Characteristic words by *gender* contextual variable without lemmatization.

are most correlated (which appear together in the same document), without taking into account the frequency with which they occur.

For example, Table 1.5 shows some results of extracting a set of words (in this case two) that have a positive and high correlation, performing the analysis for each category of a contextual variable (gender, in this case). It should be noted that the Term1 and Term2 columns do not indicate that one word precedes the other in the text. In fact, it depends on the alphabetical order.

Despite the low frequency of two terms, a notable degree of correlation can be observed between them when they appear together.

For men, the highest positive correlation is found between the words *hygiene*, and *personal* (see Table 1.5 for the correlation values). The pair *base* and *hygiene* is in the fourth position, followed by *base*, and *personal* in the fifth. *Job* is important (*know*, and *have*), and also *money nutrition*. *Sport* is related to *smoke* and *little exercise*.

Women point out *drink alcohol* and *little food* as material aspects, and psychical as *know depression*, *cheerful happy*, and *see child* in a *day healthy*.

There are some expressions that appear in both genders when conducting this study independently. It is important to note that, in the LexChar function, a word that is characterized positively because it has a higher than expected frequency in men means a negative characterization for women because it has a lower than expected frequency.

Table 1.5: 15 Highest correlations for two words using *gender* variable.

	Man			Woman		
	Term1	Term2	Correlation	Term1	Term2	Correlation
1	hygiene	personal	.8860	care	take	.8234
2	mental	physical	.7901	mental	physical	.6902
3	care	take	.7896	drink	smoke	.6900
4	base	hygiene	.7389	child	see	.6622
5	base	personal	.6785	alcohol	drink	.6535
6	drink	smoke	.6751	feel	good	.6526
7	able	perform	.6729	alcohol	smoke	.6322
8	money	nutrition	.5899	much	think	.5582
9	main	thing	.5622	physical	psychic	.5433
10	have	job	.5470	important	thing	.5283
11	difficult	important	.5444	depression	know	.5269
12	smoke	sport	.5284	day	healthy	.5141
13	difficult	moment	.5027	cheerful	happy	.5138
14	exercise	little	.4927	food	little	.4341
15	job	know	.4819	day	find	.4330

These common expressions are: *drink smoke*, *mental physical*, and *take care*.

Figure 1.10 represents nouns, adjectives, and verbs in two graphs, where the size of the correlation between words can be observed by the width of the lines connecting them.

In men, *alcohol* is related with *drink*, *smoke* and *eat* within a *normal* lifetime with *do*, *sport*, and *little exercise*. In the same way, it is important for men to be in shape (*fit* and *condition*) both *physical*, and *mental*. The *work* is close to the *money*, *able perform*, and *feel*. As in the correlation table, the words *hygiene*, *personal*, and *base* are in close proximity.

In women, *alcohol* seems to be associated only with *drink* and *smoke*, not with *sport* or *exercise*. They believe that *health* is (*be*) the *main important thing* in the *world*. They indicate words related to emotional states (*depression*, *cheerful*, and *happy*), concerned (*see* and *job*) about the *family* and *children*, as well as the *job* of all of them. In the male population, (*job* are not associated with words related to *family* or *children*. Word *nutrition* for women is related to *much*, *more*, *normal*, *little*, *eat*, *food*, *life*, and *live*. Word *money* is related to *have*, *better* and *family* (not to *job*) for women. For men, *money* is associated with *work*, *feel*, *good*, and *nutrition*.

For both groups, the word combination *take care* appears strongly related, as does *feel good*, which expresses an emotion (except in Table 1.5 for men).

Thus, the analysis of textual data, in this case responses to an open-ended question on the concept of health, provides interesting results that are enriched by crossing the textual variable with contextual variables, allowing a deeper analysis of the problem under study. It should also be noted that analyzing

Figure 1.10: Nouns, adjectives and verbs from the lemmatized text by gender. Fruchterman-Reingold algorithm.

textual data based on characterizing isolated words raises the problem of losing context when interpretations are made.

This contextual analysis of a lemmatized text can also be carried out to study, from a linguistic point of view, the different uses of grammatical categories for different values or categories of contextual variables.

Just as identifying characteristic words helps us to understand the differences between parts, characteristic documents can be considered representative examples of parts. Representative documents, whether sentences or responses, are transcribed directly from the text. They are representative of a specific part of the corpus. These documents are obtained without coding or prior interpretation. They usually include all the words, even those removed after applying a frequency-based selection threshold.

The LexChar function of the **Xplortext** package provides the characteristic words of each part (CharWord function) and the original texts of the characteristic responses that can be used as representative documents or examples of the part. This allows users to return to the context (CharDoc function).

Table 1.6 provides the five most representative responses for men and women.

1.9 Conclusion

In this chapter, we addressed the concept of contextual data from a broad perspective, covering two key aspects: the relationships between words within

Table 1.6: Characteristic responses by *gender* contextual variable.

MAN

Rank	NDoc	Criterion	Text
1	30	1.13	health of the body, hygiene, to do exercise and work
2	50	0.92	the person who eats, sleeps, works, lives a normal life
3	22	0.89	to feel good, to be physically well. not to have medical problems
4	22	0.86	health is one of the best things that a person can have. eats a good nutrition and takes care of oneself, not abusing many things, alcohol, tobacco, and many other things
5	82	0.76	physical fit of the person in which he feels no discomfort, no disease and no discomfort

WOMAN

Rank	NDoc	Criterion	Text
1	108	3.18	physical and mental wellbeing, general wellbeing of a person
2	10	2.94	health is the best thing in the world, it is an essential thing to live with happiness. there are two kinds of health, physical and mental health
3	61	2.94	physical wellbeing, I would add nothing. that the whole environment, the city helps you to wellbeing, of course, because with the noise we are going to go crazy
4	380	2.74	to be healthy, not have any illness and to live as well as possible outdoors, so as not to be sick
5	411	2.74	the best thing there is, having health you can cover everything. having health and a job is the main thing. physical fit of the person in which he feels no discomfort, no disease and no discomfort

texts, and the relationships between documents and contextual data. In both cases, it is necessary to perform corpus preprocessing, which has been facilitated in recent years with programs for spell checking, replacement of abbreviations, contractions, symbols, and named entity recognition (NER).

Traditional analysis using lexical words (graphical forms) can now also be performed on lemmatized corpora, thanks to packages that support multiple languages and produce lemmatized text outputs in formats easily exportable for processing with other programs.

Even within the same program, the objectives may require data storage in different formats for processing. A summary of these formats has been included in the Section 1.5 on vectorization.

The first aspect of contextual analysis covers the different measures and representations of the relationship between words within a text. This task generally falls within the field of Natural Language Processing. We highlighted that words within documents must be interpreted within the context of the document as a whole. Therefore, understanding the relationships between words is a fundamental process for discovering structures and interpreting results.

A survey with the question *What does health mean to you?* and gender as one qualitative closed-ended response has been used as an example within the Sections of this chapter, sometimes using graphical forms and sometimes lemmas.

The analysis of the results provided interesting findings that summarize the responses of the participants. Quite a few differences have been detected when the analysis is based on the frequency of words or when the input data has been weighted to identify the structure.

The second aspect examines how each document relates to the qualitative and quantitative contextual data that can be extracted from it. Specifically, a statistical approach using characteristic words and documents has been employed to analyze the differences in responses between men and women.

Sometimes the goal may be to measure the association between vocabulary or words and qualitative or quantitative variables, in an attempt to determine the specific topics in each category.

Global themes in a set of documents are related to the frequency of each word, either because the word is overrepresented or underrepresented. However, it is also possible to determine which documents characterize each category.

One of the challenges we faced in certain techniques is the necessity to parameterize many arguments and the issue of how small variations can lead to different configurations. Given the potential instability, it is essential to interpret the results with caution and conduct a thorough study.

2

SVD-based methods in textual analysis. An overview

2.1 Introduction

This chapter examines two widely used text-oriented machine learning methods: latent semantic analysis (LSA) and correspondence analysis (CA). Both methods place documents and words in semantic spaces calculated using the DTM described in the previous chapter. The primary goal is to identify underlying structural similarities within a given corpus of documents. Documents are examined based on their lexical content. Words are considered from the perspective of the documents in which they appear. The proximity of documents and words, as well as their relationships, is represented in reduced-dimensional spaces. In these spaces, the more two documents use the same words, the closer to each other they are, the more two words are used in the same documents, the closer to each other they are. First, in Section 2.2, we present a short example that will be referenced throughout the chapter. Then, Section 2.3 explains the principles of singular value decomposition (SVD) and eigenvalue decomposition. These are two mathematical methods used to compute the two reduced-dimensional spaces onto which the word cloud and document cloud are projected, respectively. Section 2.4 compares LSA and CA. To better understand these methods, we apply them to the short example. Section 2.5 ends the chapter with some concluding remarks.

2.2 Example

2.2.1 Verbalization task in sensometrics

This example belongs to the field of sensometry, which deals with the statistical analysis of data obtained from the sensory evaluation of products. Typically, this involves food products. Verbalization tasks are increasingly common in this field (Bécue-Bertaut et al., 2003).

DOI: 10.1201/9781003397472-2

Specifically, this example is a corpus of verbal data collected during a hall test session organized with wine producers. This corpus is concise yet authentic.

2.2.2 Data

Eight red wines from Catalonia were selected according to a full factorial design based on three factors, each with two levels:

- Two DO (designation of origin) wines: Priorat and Empordà (labeled P and E).
- Two dominant varieties: Grenache (G) and Samso (S), also known as Carignan.
- Two vintages: 2005 (05) and 2006 (06).

Wines are identified by an acronym that summarizes the three factors. For instance, a Priorat DO wine made from Grenache grapes harvested in 2005 would be labeled **PG05**. Fifteen judges, including wine experts, oenologists, sommeliers, and salespeople, tasted the wines blindly. The chosen tasting method was the *labeled sorting task*. The judges were asked to group the wines based on their sensory perceptions of the products' similarities and differences. However, the criteria used, which varied from judge to judge, are not specified. To identify these criteria, the judges were asked to describe each group with words of their choosing. We analyzed these descriptions to identify and characterize the differences between the wines. Additionally, to determine how much pleasure or displeasure the wines evoked, the judges assigned a score from one to ten, indicating how much they liked each wine. We calculated the average score for each wine, taking all judges into account, to create a quantitative variable.

2.2.3 Encoding the data

A database was created with eight rows (one for each wine) and fifteen columns (one for each judge). At the intersection of a wine row and a judge column, the free comments given by this judge for this wine are copied verbatim. The R package **Xplortext** can compile all comments related to the same wine into a single document. Thus, we have a corpus of eight documents, each named after the corresponding wine's acronym.

The corpus of all comments contains 410 occurrences of 85 different words. We eliminated words used fewer than six times, as well as the words *mature*, *long*, *spicy*, and *velvety* which were used indiscriminately among the wines. Our goal is to create a small yet substantial corpus that makes the results easy to read. After selecting the words, the remaining corpus contains 210 occurrences of 15 different words.

Table 2.1: DTM and its margins. Raw counts. Active elements only.

	acetal	acidic	animal	balanced	candy	dry	fresh	fruity	powerful	ripe	supple	tannic	unwooded	vanilla	woody	Col. marg.
PG05	1	1	1	0	1	1	2	8	1	4	2	2	1	1	1	27
PG06	0	0	0	3	0	0	3	11	2	3	2	3	1	0	2	30
EG05	1	2	3	0	1	3	0	3	1	0	0	5	1	1	3	24
EG06	0	0	1	1	2	1	1	2	1	0	0	4	1	2	8	24
PS05	3	2	1	1	0	3	2	8	1	1	0	4	2	0	0	28
PS06	0	0	0	0	2	1	0	0	2	0	1	2	0	2	10	20
ES05	0	0	0	1	3	0	1	0	2	0	0	5	0	2	11	25
ES06	2	3	3	0	1	3	2	4	2	2	1	6	0	1	2	32
Row margin	7	8	9	6	10	12	11	36	12	10	6	31	6	9	37	210

The **Xplortext** package performs these counts and constructs a DTM with I=8 documents (rows) and J=15 words (columns). This matrix, shown in Table 2.1, contains the *frequency* with which each word occurs in each document (here, wine). It takes into account the comments of all judges. For example, **PG05** is described as *fruity* 8 times. The *column margin* indicates the document's length. Document **ES06** is the longest, with 32 occurrences. The *row margin* shows how often words appear. The most frequent word is *woody* which appears 37 times. It is followed by *fruity* (36 times) and *tannic* (31 times).

2.2.3.1 Supplementary row documents and column words

We often use supplementary documents and/or words. For example, we may be interested in comments on wines of the same type produced ten years earlier or in a different region. We can also consider words that were eliminated because they were too rare but whose meaning we would like to understand better in the given context. One way to do so is to give them a supplementary role and to look for active words close to them on the plane.

In this case, we created two documents artificially. These documents are given a supplementary role without interfering with the calculation of the representation spaces. The first document, **SU05**, receives half of the comments attributed to wine **PG05**, while the other half come from wine **PS06**. These are two very different wines. In short, one is fruity, while the other is woody. **SU05** is expected to be located at the approximate center of gravity (CoG) of the two documents. The second document, **SU06**, uses the same words and frequencies as **ES05**, except *woody* is replaced by *vanilla* or *candy*. These aromas are often associated with woody wines and are often used as synonyms for *woody*, as is the case in the present data. With the help of this supplementary document, we aim to evaluate each method's ability to bring documents closer together, even if they are not described by the same words, but rather by synonyms. Regarding the supplementary words, we chose *astringent, light*, and *toasted*, all used in the comments, but less frequently than the elimination threshold of five.

2.2.3.2 Liking-score as a supplementary quantitative variable

In addition, the *liking score* described above is used as a supplementary quantitative variable. Using this type of variable has proven useful for interpreting results.

2.3 Calculation of the representation spaces

2.3.1 Introduction

We want to compare documents based on their word usage and compare words based on how they are used in the texts under consideration. Additionally, we want to explore their interrelationships. We can use the raw DTM or preprocess it to reflect our understanding of similarity. For example, we could convert the raw counts into proportions relative to document length to analyze lexical profiles or word content independent of document length. Other preprocessing methods can be considered as well, such as converting word counts to their logarithms to smooth out large differences. This choice is closely related to the method, as we will see.

Next, we need to compute the reduced-dimensional representation spaces for the document and word clouds. This allows us to project the clouds onto these spaces while preserving dispersion and eliminating noise. This process reveals existing structures in the data. Then, we will examine how these spaces are computed and their properties. The most important property is the duality between the document and word representation spaces. This duality allows us to transition between the two spaces and study the relationships between documents and words.

2.3.2 Cloud of documents and cloud of words

Let $\mathbf{Y}$ be the $I \times J$ matrix to analyze, either the raw DTM or the one resulting from some preprocessing. Let $N(I)$ be the cloud of I documents. Each point document i is placed in R^J by its coordinate vector $(y_{ij}, j = 1, \ldots, J)$. The document weights are stored in the diagonal matrix $\mathbf{D_I}$ with the generic term d_{I_i}, $i = 1, \ldots, I$. This space has a Euclidean metric $\mathbf{D_J}$, usually diagonal, whose generic term D_{J_j}, $j = 1, \ldots, J$ corresponds to the importance or weight of word j in computing the distance between documents i and i'. So,

$$d^2_{D_J}(i, i') = \sum_{j=1}^{J} (y_{ij} - y_{i'j})^2 \, d_{J_j}.$$

Similarly, let $N(J)$ be the cloud of J words. Each point word j is placed in R^I by its coordinate vector $(y_{ij}, i = 1, \ldots, I)$. The word weights are stored in the diagonal matrix $\mathbf{D_J}$ with the generic term $d_{J_j}, j = 1, \ldots, J$. This space has a Euclidean metric $\mathbf{D_I}$, usually diagonal, whose generic term $D_{I_i}, i = 1, \ldots, I$ corresponds to the importance or weight of documents i in calculating the distance between words j and j'. So,

$$d^2_{D_I}(j, j') = \sum_{i=1}^{I} (y_{ij} - y_{ij'})^2 \, d_{I_i}.$$

Thus, a weighting system and a metric are defined for each space. The chosen metric (i.e., distance) determines the shape of the cloud. The weights determine the cloud's balance and, consequently, its CoG. Note that the metric in one space is equivalent to the weighting system in the other, and vice versa. This is the only option we are considering here.

2.3.3 Measure of the dispersion of the clouds

The measure of dispersion for each of the two clouds is inertia, which is calculated with respect to either the origin or the CoG as a reference—an important choice. Inertia is equal to the sum of the product of the weight of each point and the square of its distance from the reference point. Thus, the inertia of cloud N_I with respect to its origin is:

$$In\,(N_I) = \sum_{i=1}^{I} d_{I_i} \cdot d^2\,(i, O)\,.$$

Similarly, the inertia of the cloud N_J with respect to its origin is:

$$In\,(N_J) = \sum_{j=1}^{J} d_{J_j} \cdot d^2\,(j, O)\,.$$

Note that origin can be replaced with CoG in these formulas.

2.3.4 Projection of a cloud on an axis and its inertia

Let v be a unit vector for the metric D_J of any given axis of R^J (i.e., $v^T\mathbf{D_J}v = 1$). The coordinate vector of $N(I)$ projected onto this axis, denoted F_v, is calculated as $F_v = \mathbf{Y}\mathbf{D_J}v$. The inertia of the cloud projected onto v is equal to $\sum_{i=1}^{I} d_{I_i}[F_v(i)]^2$ or, in matrix form, $F_v^T\mathbf{D_I}F_v$. Similarly, let u be a unit vector for the metric D_I of any given axis of R^I (i.e., $u^T\mathbf{D_I}u = 1$). The coordinate vector of $N(J)$ projected onto this axis, denoted G_u, is calculated as $G_u = \mathbf{Y^T}\mathbf{D_I}u$. The inertia of the cloud projected onto u is equal to $\sum_{j=1}^{J} d_j[G_u(j)]^2$ or, in matrix form, $G_u^T\mathbf{D_J}G_u$.

2.3.5 Calculating the axes of maximum inertia using GSVD

We now search for the axis of maximum inertia of the cloud $N(I)$—that is, the unit axis v for the metric $\mathbf{D_J}$ such that $F_v^T\mathbf{D_I}F_v$ is maximum. We find this axis using generalized singular value decomposition (GSVD), which we explain in the following.

GSVD states that any $I \times J$ matrix $\mathbf{Y}$ of rank $S, S \leq (\min\{I, J\})$, can be decomposed into the product of three matrices denoted $\mathbf{U}$, $\mathbf{\Sigma}$ and $\mathbf{V^T}$, namely

$$\mathbf{Y} = \mathbf{U \Sigma V^T}$$

where $\mathbf{U}$ is an $I \times S$ matrix, $\mathbf{V}$ is an $J \times S$ matrix, and $\mathbf{\Sigma}$ is an $S \times S$ diagonal matrix. The $\mathbf{U}$ columns, called *left singular vectors*, are $\mathbf{D_I}$-orthonormal, i.e. $\mathbf{U^T D_I U} = \mathbf{Id}$. The $\mathbf{V}$ columns, called *right singular vectors*, are $\mathbf{D_J}$-orthonormal, i.e. $\mathbf{V^T D_J V} = \mathbf{Id}$. The diagonal elements of $\mathbf{\Sigma}$ are called the singular values of $\mathbf{Y}$. The GSVD decomposition is unique up to sign permutations. By convention, the singular values are all positive and in descending order. The vector columns u_s, $s = 1, \ldots, S$ of $\mathbf{U}$ and the vector columns v_s, $s = 1, \ldots, S$ of $\mathbf{V}$ are arranged in the same order as the singular values σ_s, $s = 1, \ldots, S$ of the same rank.

The vector v_1, i.e. the one associated with the largest singular value σ_1, is such that the inertia of the document cloud projected on this axis is maximum and equal to σ_1^2. The axis orthogonal to v_1, such that the corresponding plane maximizes the inertia of the cloud projected on a plane, is the singular vector associated with the second singular value, that is, v_2. And so on for following axes. The sequence of $\mathbf{D_J}$-orthonormal axes that maximizes the projected inertia is thus defined by the sequence of S right singular vectors of $\mathbf{Y}$ ordered by decreasing singular values, assumed to be distinct. Finding the plane of maximum inertia, or instead finding the first axis of maximum inertia followed by the second, leads to the same solution. More generally, the s dimensional hyperplane of maximum inertia contains the corresponding $s - 1$ dimensional one, and so on down to the plane and then the axis of maximum inertia. That is, the subspaces of maximum inertia, also known as *factorial subspaces*, are nested. Note that the S vector columns $\mathbf{V}$ form together a $\mathbf{D_J}$-orthonormal basis of the space R^J (see Figure 2.1).

Symmetrically, the left singular vectors u_s, $s = 1, \ldots, S$ form a $\mathbf{D_I}$-orthonormal basis of the space R^I. They are the unit axes of maximum inertia of the cloud $N(J)$ for the metric $\mathbf{D_I}$. Again, these axes are also ordered by decreasing inertia, according to the associated singular value. We are not going to go into more detail.

Note that the GSVD method computes the maximum inertia axes of the two clouds simultaneously.

2.3.6 Calculating the axes of maximum inertia using eigendecomposition

SVD and GSVD are closely related to the eigendecomposition of a square symmetric matrix. With the same goal in mind—finding the axes of maximum inertia of both clouds, $N(I)$ and $N(J)$—this can be achieved by successively calculating the eigenvectors and eigenvalues of the matrices $\mathbf{Y^T D_I Y D_J}$ and

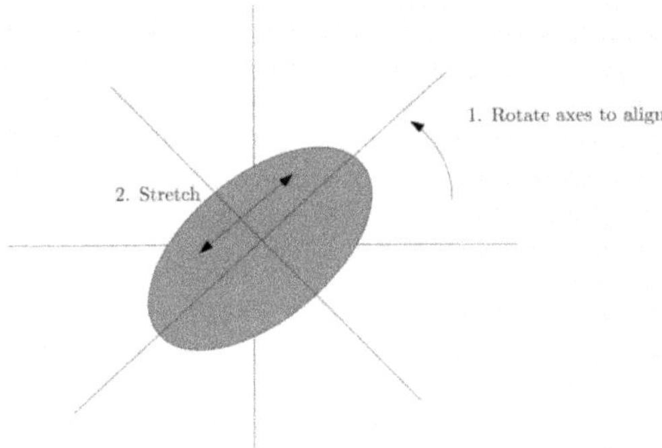

Figure 2.1: New orthogonal axes.

$\mathbf{Y}\mathbf{D_J}\mathbf{Y}^\mathbf{T}\mathbf{D_I}$. The obtained eigenvectors are equal to the right and left singular vectors, $\mathbf{V}$ and $\mathbf{U}$, respectively, calculated by GSVD. They respect the same orthonormality constraints as in GSVD: $\mathbf{U}^\mathbf{T}\mathbf{D_I}\mathbf{U} = \mathbf{Id}$ and $\mathbf{V}^\mathbf{T}\mathbf{D_J}\mathbf{V} = \mathbf{Id}$. In this case, the eigenvalues λ_s, $s = 1, \ldots, S$, which are stored in the matrix $\mathbf{\Lambda}$ and are common to both eigendecompositions, are equal to the squares of the singular values. In practice, only one eigendecomposition is necessary. The results of the second decomposition can be obtained using the transition relationships presented below. The decomposition of the lowest-dimensional matrix is selected because it is the least time-consuming.

For our purposes, the results are equivalent regardless of the calculation method chosen: GSVD or eigendecomposition. Note that matrix $\mathbf{Y}$, which is to be decomposed, may have resulted from previous DTM transformations. As we will see in LSA and CA, it is possible to perform centering, standardization, and other modifications.

The concrete calculation is completely determined by the triplet formed by the matrix to be decomposed, noted $\mathbf{Y}$ here, the metric in the row space $\mathbf{D_J}$ (= weighting system in the column space) and the metric in the column space $\mathbf{D_I}$ (= weighting system in the row space). We will use the notation GSVD($\mathbf{Y}$, $\mathbf{D_J}$, $\mathbf{D_I}$) as well as *general analysis* (GA) ($\mathbf{Y}$, $\mathbf{D_J}$, $\mathbf{D_I}$), the latter in order to follow the terminology employed by Lebart et al. (2006). Occasionally, we will also use the notation PCA($\mathbf{Y}$, $\mathbf{D_J}$, $\mathbf{D_I}$), where PCA stands for *principal component analysis*.

All SVD-based methods, also known as factorial methods, use the same process to calculate the axes of inertia and the eigenvalues or singular values. The following properties are common to all of these methods. Standard factorial methods, such as PCA, CA, multiple correspondence analysis (MCA) and so on, can be viewed as generalized singular value decompositions (GSVDs) or General Analyses (GAs) provided the necessary transformations are applied to the matrices. We will exploit this possibility in the following chapters.

2.3.6.1 Row and column factors

Regardless of the method used to calculate the axes of inertia, the coordinate vectors of the documents and words on them are then sought. They are called *factors on the rows* and *factors on the columns*, respectively. The factor on the rows of rank s is calculated as the projection of the cloud $N(I)$ on axis of inertia s as follows: $\mathbf{F_s} = \mathbf{YD_J} v_s$, $s = 1, \ldots, S$. The S row factors are stored in the $I \times S$ matrix $\mathbf{F}$. Symmetrically, the factor on the columns of rank s is calculated as follows: $\mathbf{G_s} = \mathbf{Y^T D_I} u_s$, $s = 1, \ldots, S$. The S column factors are stored in the $J \times S$ matrix $\mathbf{G}$.

The *inertia of the row factor* s equal to $\sum_{i=1}^{I} d_{I_i} F_s^2(i) = \sigma_s^2 = \lambda s$. The *inertia of the column factor* s is equal to $\sum_{j=1}^{J} d_{J_j} G_s^2(j) = \sigma_s^2 = \lambda_s$. In both spaces, the factors of rank s have the same inertia, which is equal to both the squared singular value and the eigenvalue of the same rank.

The total inertia of the cloud $N(I)$ (resp. $N(J)$) is the sum of the inertias along the axes, i.e., the sum of all the squared singular values or the sum of all the eigenvalues. In both spaces, the percentage of inertia explained by the s-th axis is equal to the ratio of the associated squared singular value to the total inertia, i.e., $\frac{\sigma_s^2}{In(I)} = \frac{\sigma_s^2}{In(J)}$. The percentage of inertia explained by a factorial plane equals the ratio of the sum of the squared singular values associated with its two axes to the total inertia. Typically, the sum of the percentages of inertia explained by the axes indicates the quality of representation in a plane or higher-dimensional subspace.

2.3.6.2 Transition formulas

It is easy to verify that $\mathbf{F} = \mathbf{U}\Sigma$ and $\mathbf{G} = \mathbf{V}\Sigma$. In other words, the factors in one space are collinear with the singular vectors in the other. Both clouds exist in spaces with the same number of dimensions and have the same number of factorial axes with non-zero inertia. These transition relationships reflect our two-sided approach to DTMs: document-based and word-based. They allow us to calculate row factors as a function of column factors, and vice versa. Thus,

$$\mathbf{F_s} = \frac{1}{\sqrt{\lambda_s}} \mathbf{YD_J G_s}$$

and

$$\mathbf{G_s} = \frac{1}{\sqrt{\lambda_s}} \mathbf{Y^T D_I F_s}.$$

2.3.6.3 Supplementary rows and columns

As mentioned above, supplementary elements, documents, and/or words are useful for interpretation. The coordinates of these elements can easily be calculated using transition relationships. We will see their usefulness in the running example.

Note that both quantitative and qualitative variables can serve as supplementary variables. In this example, the liking score is a quantitative supplementary variable placed on the axes according to its correlation with them. Supplementary categories are placed on the axes as if they were supplementary documents and are assigned a weight equal to the total weight of their documents.

2.4 Comparison of latent semantic analysis and correspondence analysis

2.4.1 Introduction

Latent Semantic Analysis (LSA), also known as Latent Semantic Indexing (LSI), emerged in the field of information retrieval in the late 1980s (Deerwester et al., 1990).

LSA is an SVD-based method that retrieves relevant documents for a query, even if the query contains synonyms instead of the exact terms used in the documents. There are many variants, depending on how the DTM is preprocessed. All variants provide a spatial configuration of documents and words in a reduced-dimensional space. This space is referred to as a *semantic space* in the cited paper. In this space, document retrieval and machine learning methods occur, including topic modeling, text clustering, text categorization, and authorship attribution.

CA, the second method discussed here, originated in statistics in the 1960s, predating LSA. It was primarily developed by Escofier [Cordier] (1965) and Benzécri (1973, 1981). The latter gave the method its current name and conceptualized CA as a tool for addressing linguistic issues in accordance with distributional linguistics (Harris, 1954). CA is an SVD-based method that begins with a preprocessed DTM. From the outset, the method focuses on visualizing the proximity of documents and words, as well as their interrelationships, in a reduced-dimensional space. CA is a versatile method that can be used for objectives similar to those of LSA. It can also be used for a more exploratory approach, as well as for addressing other concerns. Although CA has many applications in textual analysis, it has also been used with classical categorical data. This may have prevented the method from being more widely adopted in natural language processing (NLP).

2.4.2 LSA

2.4.2.1 Preprocessing of the DTM and LSA variants

LSA offers several variants for preprocessing the lexical table. Qi et al. (2024) provides a detailed and well-argued comparison of LSA and CA, considering four variants. We briefly present these variants below and examine one of them in more detail.

The variants apply singular value decomposition (SVD) to either the raw DTM or the preprocessed tables. Preprocessing mitigates some of the challenges associated with analyzing the raw DTM. We examine the same variants as in the cited paper: the raw DTM, two row normalization options (LSA-L1 and LSA-L2), and an option named Term Frequency $\times$ Inverse of Document Frequency (LSA TF-IDF). The corresponding calculations are detailed below.

In LSA-L1, row normalization involves dividing each element in a row by the sum of that row. In other words, each row is replaced by its profile. Each document i is represented by a point whose coordinates are the elements of its profile, $y_{ij}/y_{i\cdot}, j = 1, \ldots, J$. Consequently, the sum of each row is equal to one. This cancels out the effect of differences in document length, which can be significant when lengths vary greatly.

In LSA-L2, row normalization involves dividing each element of a row by the sum of its squared elements. Thus, each document is represented by a point whose coordinates are $y_{ij}/\sqrt{\sum_{j=1}^{J} y_{ij}^2}$, $j = 1, \ldots, J$. This normalization has a similar objective to LSA-L1. Additionally, it attempts to limit the impact of high-frequency terms.

In the case of the latter preprocessing proposal, TF-IDF, SVD is applied to the table with the generic term $y_{ij}\left(1 + log\left(\frac{I}{I_j}\right)\right)$, $i = 1, \ldots, I$, $j = 1, \ldots, J$, where I_j is the number of documents containing the word j. Then, the log of the inverse proportion of documents containing term j (i.e., $log\left(\frac{I}{I_j}\right)$) was added to the term's frequency (TF), $(=y_{ij})$. This emphasizes the role of specialized terms, making it easier to categorize documents by topic based on these terms.

Note that this definition of TF-IDF differs slightly from the one discussed in Chapter 1 and the one commented on by Lebart et al. (2019), both of which differ from one another. In fact, there are several forms of TF-IDF. For the purposes of this chapter, we adopt the definition from Qi et al. (2024). This allows readers to compare TF-IDF and CA results using the same definition of TF-IDF for two different examples. Note that the CA definition does not vary.

2.4.2.2 Metrics used in LSA

All LSA variants use identity metrics/uniform weighting systems. Thus, LSA is the GSVD defined by the triplet ($\mathbf{Y}$, $\mathbf{Id}$, $\mathbf{Id}$). Alternatively, it is an SVD applied to the preprocessed matrix $\mathbf{Y}$.

2.4.2.3 Important common characteristics

All LSA variants share the following key characteristics:

- In the document (=row) space, the document weights are uniform ($\mathbf{D_I} = \mathbf{Id}$) and the metric is the classical Euclidean distance ($\mathbf{D_J} = \mathbf{Id}$). In the word (=column) space, the weights are uniform as well (i.e., $\mathbf{D_J} = \mathbf{Id}$), and the metric is the classical Euclidean distance ($\mathbf{D_I} = \mathbf{Id}$). The weighting system in the row space corresponds to the metric in the column space, and vice versa. Since neither the row nor the column cloud is centered, the inertia in both spaces is calculated relative to the origin rather than the CoG of either cloud. Consequently, in each space, the first axis captures variability from the origin to CoG of the cloud. This primarily shows size differences, which are rarely interesting. Most importantly, since the inertia axes must be orthogonal, subsequent axes cannot point in significant dispersion directions (see Figure 2.2).

- The inertia retained on the first factorial axis does not make much sense. To generate some ideas, let's experiment with the example. Suppose each judge adds the same word, such as *wine*, to each wine's description. Then, observe how the inertia varies. This new word would appear 120 times, describing each wine the same way in terms of absolute frequency. Therefore, it should not be considered important. However, in the case of LSA-RAW, introducing this word increases total inertia significantly, from 938 to 2,738, primarily due to a substantial increase along the first axis. Consequently, the proportion of inertia retained by the first axis increases from 62.28% to 86.88%, while the proportion retained by the second axis decreases from 28.16% to 9.65%. Meanwhile, the configuration visualized on the (2,3) plane remains practically unchanged. Similar variations occur for the other variants. These results clearly indicate that the first axis is not of interest.

- In LSA, neither total inertia nor inertia along an axis can be interpreted based on reference values.

- Since the analyzed matrix is nonnegative, the coordinates have the same sign on the first factorial axis for both documents and words. The longer the document or the more frequently the word appears, the larger its absolute coordinate tends to be. On this axis, the row and column coordinates are roughly proportional to their length, which is not very informative.

- LSA allows you to compare documents or words to each other and measure their association. To accomplish this, we calculate the cosine distance or dot product between the coordinate vectors representing these elements in the reduced space returned by LSA.

- The notion of topics is widely used to synthesize results. In the document space, each axis of inertia can be considered a topic or composite word. The meaning of a topic is revealed by the list of words that have a high contribution or coordinate on the same rank axis in the word space. These words form a whole that acquires meaning through frequent association. These topics can be summarized by keywords, also known as meta-words, which play an important role in information retrieval. Identifying these

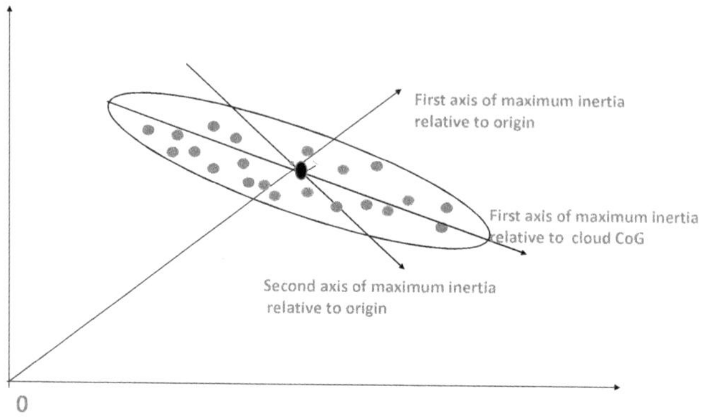

Figure 2.2: Inertia axes. Inertia relative to either origin or CoG.

topics is one of the goals of the analysis. The same word can be associated with multiple axes. As these axes are orthogonal, the word will be associated with different words on each axis. This gives the word different meanings, making it polysemous within the corpus.

2.4.2.4 LSA results from the example

The results were obtained using the script **chap2**. The **svs** package available in R CRAN is used to perform LSA. However, to get sharper graphics, we use the graphical capabilities of the package **Xplortext**. In this example, the four LSA variants produce similar results except for the singular values and amount of inertia, which vary greatly among the variants (see Table 2.2). However, the percentages of inertia retained are close when considered axis by axis. Furthermore, the document and word configurations remain stable from one variant to the next on both the (1,2) and (2,3) planes.

Regardless of the variant, the three words that appear most frequently—*woody* (37 times), *fruity* (36) and *tannic* (31)—have the most negative coordinates on the first axis. However, they are not always in the same order.

First, we examine the retained inertias on the axes and their percentage of the total inertia (see Figure 2.3, top). The values themselves are not meaningful. For example, adding ten, hundred or more instances of a word to each document can considerably increase the percentage of inertia corresponding to the first axis, even if the word does not differentiate between documents. This suggests that the first axis is not useful here. Note that the longer the document or the higher the word frequency, the farther the document is from the origin. In this case, document coordinates are less variable than word frequencies because document lengths are more consistent.

Now, let's transition to the (2,3) plane to reveal the data structure. This is the only plane in which we are interested. The second axis accounts for 26% of the inertia, placing **PS06**, **ES05**, and **EG06** on the positive side, while the other wines are on the negative side. These wines are distributed along axis 3, which accounts for 9% of the inertia. They range from **ES06** and **EG05**, which have the most negative coordinates, to **PG06**, which stands out with a strongly positive coordinate.

Table 2.2: Variants of LSA applied to the example. Singular values and inertia.

```
---------------------------------------------------------
LSA RAW
      Sing. value  Inertia  % of inertia Cum.% of inertia
dim 1      24.17    584.19   62.28           62.28
dim 2      16.25    264.18   28.16           90.44
dim 3       7.87     61.95    6.60           97.05
  Total inertia: 938

---------------------------------------------------------
LSA L1
      Sing. value  Inertia  % of inertia Cum.% of inertia
dim 1       0.95      0.90   63.36           63.36
dim 2       0.63      0.40   28.20           91.56
dim 3       0.28      0.08    5.62           97.18
  Total inertia: 1.42

---------------------------------------------------------
LSA L2
      Sing. value  Inertia  % of inertia Cum.% of inertia
dim 1       2.26      5.13   64.11           64.11
dim 2       1.43      2.05   25.69           89.79
dim 3       0.75      0.57    7.08           96.87
  Total inertia: 8

---------------------------------------------------------
LSA TF-IDF
      Sing. value  Inertia  % of inertia Cum.% of inertia
dim 1      43.47   1054.20   60.57           60.57
dim 2      21.27    452.20   25.98           86.55
dim 3      12.43    154.60    8.88           95.43
  Total inertia: 1740.50

----------------------------------------------------------
```

The words that characterize the axes help us to understand this structure in terms of wine characteristics. Since *woody* is on the positive side of the second axis with a high coordinate and *fruity* is on the negative side, this axis contrasts documents based on whether the corresponding wines are *woody* or *fruity*. The third axis contrasts the undesirable wine characteristics *acidic*, *acetaldehyd*, and *drying* with the desirable characteristics *fruity*, *ripe* and *balanced*. Thus, this axis clearly contrasts quality. Note that most of the words are close to the origin. Due to the significant frequency differences between the words, which are much larger for *woody*, *fruity* and *tannic*, than for the other words and the effect on the coordinates, this plot does not clearly illustrate the similarities or differences between them. Now, let's discuss the supplementary elements.

The supplementary document **SU05** lies near the CoG of **PG05** (fruity wine) and **PS06** (woody wine) in accordance with the fact that **SU05** shares

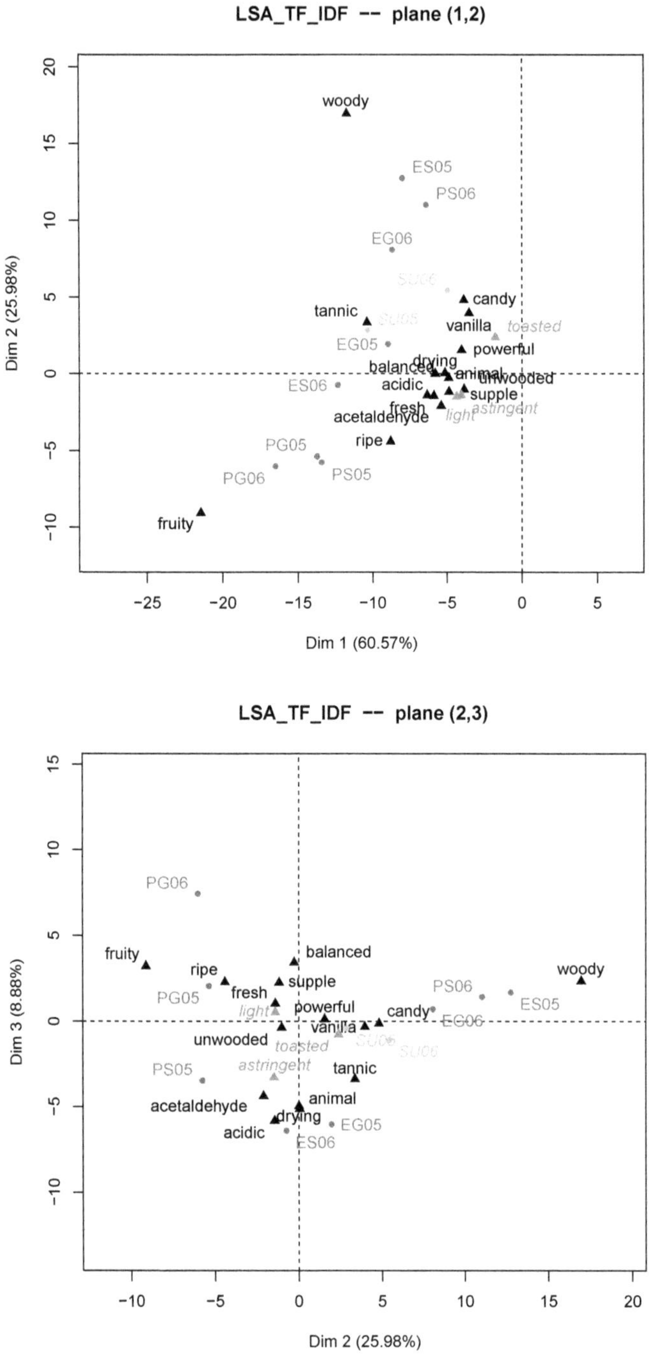

Figure 2.3: LSA TF IDF. Row and column planes. Top: (1,2) plane; bottom: (2,3) plane.

half of its comments with **PG05** and the other half with **PS06**. We note that **SU06** is closer to **EG06** than to **ES05** in terms of cosine distance.

Regarding the supplementary words, *astringent* is clearly close to the negative words (e.g., *acetaldehyde, animal, drying*) for both Euclidean and cosine distances. *Light* is close to *fresh* and *fruity* (active words) which is not surprising. *Toasted* is close to *vanilla* and *candy* but not so close to *woody*, about which little can be said. These similarities are measured by cosine distance.

Finally, we examined the liking score, which revealed a strong positive correlation with the third axis ($r = 0.8$). This confirms that this axis represents quality, as we had previously observed with the words. This information would be essential in other cases, such as when the words themselves are unclear or the study is conducted in an unknown language.

2.4.3 CA

As said above, CA is based on the distributional analysis developed by Harris (1954). Each word's distribution is defined as *the sum of all its environments or, equivalently, the set of its co-occurrences.* Harris argued that interword dependencies alone are sufficient to determine correspondences in a set of documents, regardless of meaning. Benzécri (1973, 1981) proposed CA as a means of translating distributional analysis into a statistical tool for analyzing textual data. Lebart et al. (1998) expanded upon this work by incorporating free-text responses. This led them to develop the concept of analyzing textual and contextual data together.

2.4.3.1 Preprocessing the DTM in CA

First, the corpus is encoded into a DTM, which is then transformed into a proportion matrix, **P**. The generic term p_{ij} is calculated by dividing the DTM by N, the total number of occurrences. Summing **P** per row yields the column margin with the generic term $p_{i.}$, which is equal to the relative length of document i, $i = 1, \ldots, I$. Summing **P** per column yields the *row margin* with the generic term $p_{.j}$, equal to the relative frequency of word j, $j = 1, \ldots, J$. The grand total of matrix **P** is 1. Both the DTM and **P** matrices are called *contingency tables* in the CA context.

Distributional analysis raises the question of whether the document and word sets are statistically independent. If they were, we would observe $p_{ij} = p_{i}.p_{.j}$ or, equivalently, $Np_{ij} = Np_{i}.p_{.j}$. In other words, the observed counts would equal the expected counts under the independence hypothesis. Conversely, if some cells in matrix **P** deviate from the former equality, then documents and words are associated. This leads to the definition of the association rate between document i and word j as:

$$\tau_{ij} = \frac{p_{ij}}{p_{i\cdot} \cdot p_{\cdot j}} = \frac{N \cdot p_{ij}}{N \cdot p_{i\cdot} \cdot p_{\cdot j}}.$$

If document i and word j attract each other, then τ_{ij} is greater than 1, and less than 1 if they repel each other. If documents and words were fully independent, the association rate would equal 1 for each document-word pair.

The coefficients τ_{ij} are stored in the $I \times J$ matrix $\mathcal{T}$, which summarizes deviation from independence and is analyzed using CA. Next, we need to define the weighting systems and distances used in the document and word spaces.

2.4.3.2 Distances and weighting systems

In CA, the distance between documents i and i' reflects their different usage of words, as shown in matrix $\mathcal{T}$, but weighted by the words' relative frequency, given by $p_{\cdot j}$, $\{j = 1, \ldots, J\}$. Therefore, the squared distance between documents i and i' is given by:

$$d^2(i,i') = \sum_{j=1}^{J} p_{\cdot j} \left(\left(\frac{p_{ij}}{p_{i\cdot} \cdot p_{\cdot j}} \right) - \left(\frac{p_{i'j}}{p_{i'\cdot} \cdot p_{\cdot j}} \right) \right)^2 .$$

This can be rewritten as follows:

$$d^2(i,i') = \sum_{j=1}^{J} \frac{1}{p_{\cdot j}} \left(\frac{p_{ij}}{p_{i\cdot}} - \frac{p_{i'j}}{p_{i'\cdot}} \right)^2 .$$

Symmetrically, the squared distance between words j and j' is given by:

$$d^2(j,j') = \sum_{i=1}^{I} p_{i\cdot} \left(\left(\frac{p_{ij}}{p_{\cdot j} \cdot p_{i\cdot}} \right) - \left(\frac{p_{ij'}}{p_{\cdot j'} \cdot p_{i\cdot}} \right) \right)^2 .$$

This can be rewritten as follows:

$$d^2(j,j') = \sum_{i=1}^{I} \frac{1}{p_{i\cdot}} \left(\frac{p_{ij}}{p_{\cdot j}} - \frac{p_{ij'}}{p_{\cdot j'}} \right)^2 .$$

The axes of maximum inertia are calculated by the GSVD($\mathcal{T}$, $\mathbf{D_J}$, $\mathbf{D_I}$). The diagonal matrix $\mathbf{D_I}$ has for generic term $p_{i\cdot}$. The diagonal matrix $\mathbf{D_J}$ has for generic term $p_{\cdot j}$.

Notes

- Both distances, called chi-square distances, are distances between profiles.
- We can see that the distance between two documents i and i' which use the different words with the same relative frequency, i.e. such that $\frac{p_{ij}}{p_{i.}} = \frac{p_{i'j}}{p_{i'}}$ for $\{j = 1, \ldots, J\}$, would be equal to 0.
- Similarly, if two words appear with the same frequency in different documents, i.e. such that $\frac{p_{ij}}{p_{\cdot j}} = \frac{p_{ij'}}{p_{\cdot j'}}$ for $\{i = 1, \ldots, I\}$, they have a distance equal to 0.

2.4.3.3 Important features

- CA is related to the chi-square statistic, which is used to test for independence between the rows and columns of a contingency table. The chi-square statistic summarizes the squared differences between the observed and expected counts under the assumption of independence as follows:

$$\chi^2 = \sum_{i,j} \frac{(Np_{ij} - Np_{i.} \cdot p_{.j})^2}{Np_{i.} \cdot p_{.j}} = N \sum_{i,j} \frac{(p_{ij} - p_{i.} \cdot p_{.j})^2}{p_{i.} \cdot p_{.j}}.$$

- The singular values are all nonnegative and less than or equal to 1.
- The first singular value is always equal to 1.
- Since the analyzed matrix is non-negative, the coordinates on the first factorial axis are all equal to 1 or -1.
- As in LSA, the first axis in both spaces captures variability from the origin $\mathbf{O}$ to the CoG. In this case, however, the first axis is orthogonal to the hyperplane containing the $N(I)$, respectively $N(J)$ cloud in terms of chi-square distance. Projecting any point in the cloud onto this axis results in merging with the centroid. The inertia of the $N(I)$, respectively $N(J)$, cloud projected onto this axis is equal to one, so it is not of interest. Since subsequent axes are orthogonal to the first, the analysis can be performed with respect to the origin or the CoG. Except for the first trivial axis, the other axes are the same.
- It is also possible to analyze the cloud centered on its CoG, i.e. to apply a GSVD (using the same metrics as above) to the table $\bar{\mathcal{T}}$ with the generic term $\bar{\tau}_{ij}$

$$\bar{\tau}_{ij} = \frac{p_{ij} - p_{i.} \cdot p_{.j}}{p_{i.} \cdot p_{.j}}.$$

  This results in the same inertia axes except for the first one, which no longer exists. In classical CA, even if the calculations are performed with the former off-center table, the first axis is eliminated from the results because it is not of interest.
- Several interpretation aids are defined in CA. Specifically, the words that contribute the most to a given axis are identified. This makes it possible to identify differences between the topics considered by the axes. This is one of the objectives of CA interpretation. As in LSA, the same word can be associated with several axes, each with different words. This gives the word different meanings, making it polysemous in the corpus.

- The transition relationships connecting the factors on the rows and the factors on columns on axis s are expressed as follows:

$$F_s(i) = \frac{1}{\sqrt{\lambda_s}} \sum_{j=1}^{J} \frac{p_{ij}}{p_{i.}} G_s(j),$$

$$G_s(j) = \frac{1}{\sqrt{\lambda_s}} \sum_{i=1}^{I} \frac{p_{ij}}{p_{.j}} F_s(i).$$

Therefore, on axis s, the row point i is, up to a constant, at the CoG of the column points j, and vice versa. The weights are determined by the components of the row (resp. column) profile, i (resp. j). An important result of CA is the visualization of similarities between documents and words, as well as their associations.

2.4.3.4 CA results from the example

The results were obtained using the script **chap2**. The **svd** function, available in R-CRAN, is applied to the matrix of association rates. This retains the first axis associated with the first singular value, which is always equal to one. Note that standard packages offering CA calculations omit this axis. Including it allows for a more accurate comparison with LSA. **Xplortext**'s graphical capabilities are used to create sharper graphics. The singular values and eigenvalues calculated by CA are shown in Table 2.3.

Table 2.3: CA applied to the example. Singular values and inertia.

CA

	Sing. value	Inertia	% of inertia	Cum. % of inertia
dim 1	1.00	1.00	58.93	58.93
dim 2	0.63	0.40	23.75	82.68
dim 3	0.43	0.18	10.74	93.42
Total inertia:	1.70			

Figure 2.4 shows the joint representation of documents and words on the (1,2) and (2,3) CA planes. For each cloud, the first axis connects the origin to the CoG and is orthogonal to the hyperplane containing the cloud, given the chosen distance. In any analysis, all documents and words on the first axis have the same coordinate, either -1 or $+1$; here, it is -1. Typically, this axis is discarded, and the inertias of the remaining factors are recalculated. In this case, only the (2,3) plane is of interest.

The configuration of the documents on this plane resembles that on the LSA (2,3) plane. On axis 2, which retains 57.82% of the inertia when the first axis is disregarded, an opposition emerges between the wines **PS06**, **ES05**,

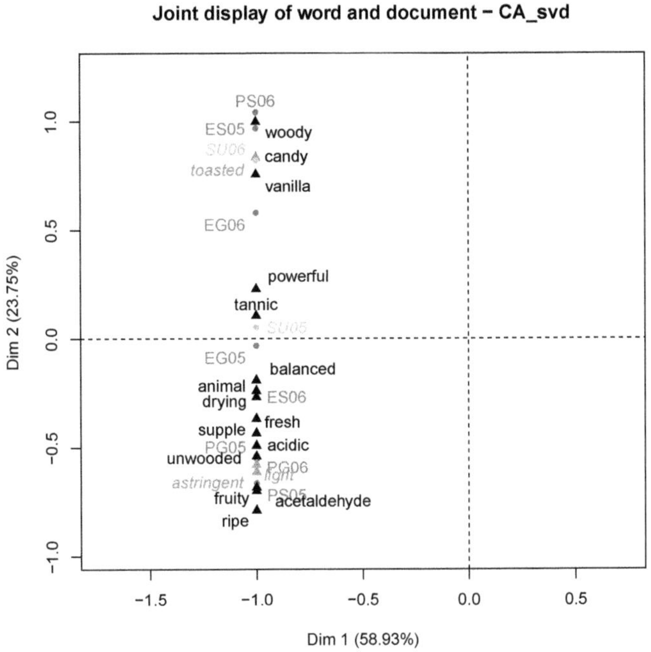

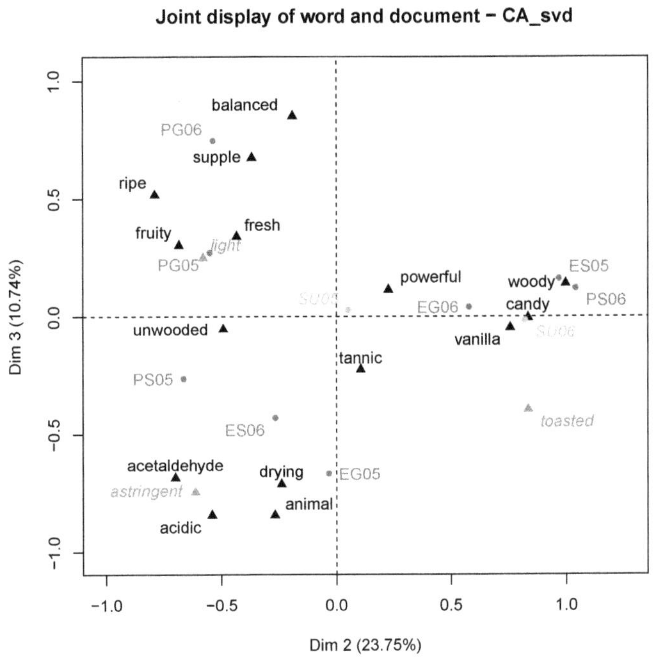

Figure 2.4: CA. Row and column visualization. Top:(1,2) plane; bottom: (2,3) plane.

EG06, and the others. These three wines are located to the right of the axis and form a compact group. Axis 3, which accounts for 24.14% of the inertia, distributes the remaining five wines—**EG05**, **ES06**, **PS05**, **PG05**, and **PG06**—along the axis in that order. The remaining three wines remain close to the CoG. Calculating confidence ellipses around both documents and words (Efron, 1979; Lebart et al., 2006)—a common CA practice— confirms these observations. Figure 2.5 shows that the confidence ellipses around the first three wines overlap but do not overlap with the other five. This indicates that the first three wines are similar to each other, yet different from the last five. This figure also shows a slow, continuous evolution of differences between the last five wines.

Now, let us examine the words. Compared to the LSA configuration (TF-IDF variant), this one is much clearer. The three most frequent words (*woody*, *fruity* and *tannic*) do not have particularly high coordinates. The other words are also not found near the origin or CoG of the cloud. This is because, in CA, it is a word's profile—how it is distributed across documents—that matters. Additionally, due to the way transition relationships are expressed in CA, words are at the CoG of the documents that use them, except for one constant.

Along axis 2, on the positive side, we see *woody*, *vanilla* and *candy*. In contrast, *unwooded* and other words form three groups along axis 3, indicating either defective characteristics, such as *animal, acid, drying, acetaldehyde*, on the negative side of axis 3, or fruity characteristics, such as *fruity, light, fresh* and *ripe*, with low-to-medium positive coordinates. Finally, qualities of good wine, such as *supple* and *balanced*, have high positive coordinates. Thus, the second axis contrasts woody and non-woody wines, and the third axis ranges wines by quality. On this axis, the more imperfect a wine is, the more negative its coordinate. The more positive the coordinate, the more highly regarded the wine.

Regarding supplementary elements, **SU06** is closer to **ES05** than to the other wines, though only by a small margin compared to its distances from **PS06** and **EG06**.

The chi-square distance is used here. It supports the hypothesis that documents described by *woody* or distributionally synonymous words, such as *vanilla* and *candy*, should be close to each other in the factorial plane. As expected, document **SU05** is close to the CoG between **PG05** and **PS06**. Regarding the words, *astringent* is close to negative terms, and *light* is between *fruity* and *fresh*, which is not surprising. In fact, the tendency of a word to be negative or positive can be determined by observing the words to which it is close. *Toasted* is positioned between woody aromas and words indicating defects. It is used to describe two woody wines and one defective wine. However, due to its low frequency, we cannot derive any information from it.

Finally, we use the liking score as supplementary quantitative variable. It is negatively correlated with the second axis at -0.16 and positively correlated with the third axis at 0.83. Therefore, perceived quality has little to do with

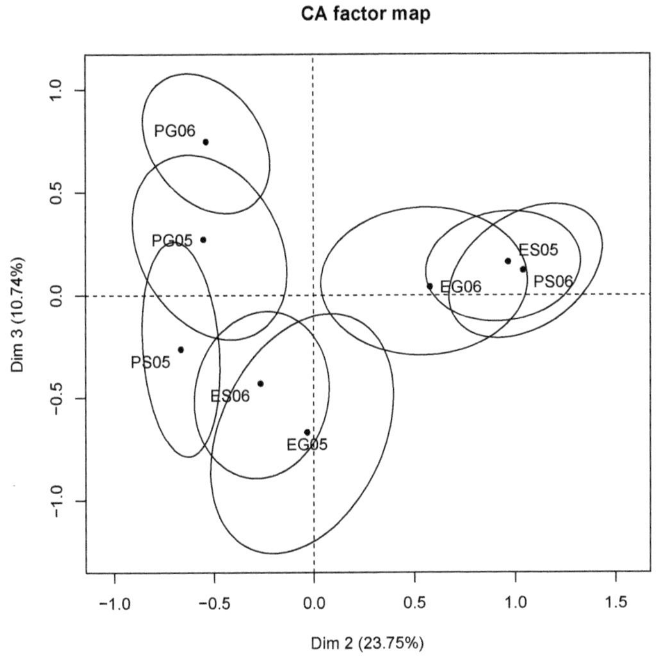

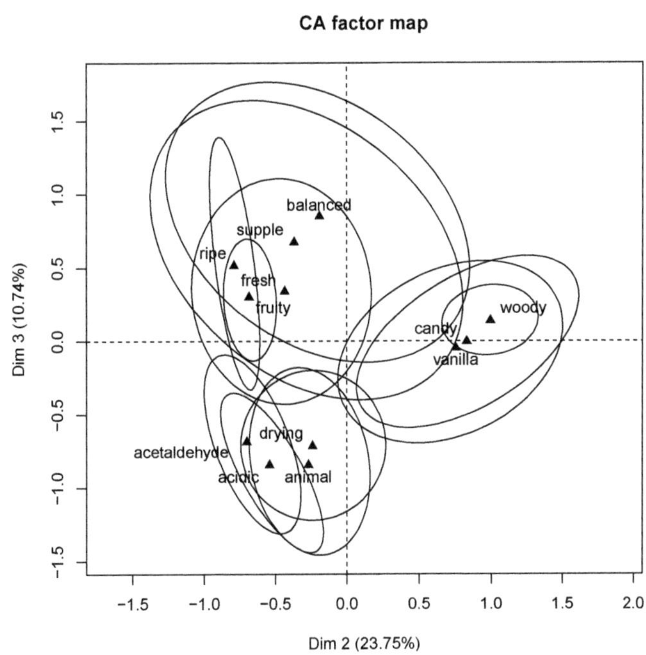

Figure 2.5: CA. Confidence ellipses on the (2,3) plane. Top: documents; bottom: words.

the second axis, which contrasts *woody* wines with *non-woody* wines. This is consistent with the contrast on the third axis between poor-quality wines, which are *acidic* and have an *animal* scent, and good-quality wines, which are *balanced* and *supple*. These results align with the information from the liking score. These qualities can be found in wine **PG06** and, to a lesser extent, in wine **PG05**. Defects, especially acidic and animal scents, are often associated with **EG05**.

These results are similar to those observed with LSA (TF-IDF variant), but more precise, especially with regard to words.

2.5 Conclusion

Both LSA and CA provide representations of document and word similarities, as well as their associations. However, LSA uses cosine distance, whereas CA uses Euclidean distance, which is more easily perceived by the human eye. Additionally, unlike CA, LSA searches for dispersion axes with respect to the origin rather than the CoG of the clouds. This results in more blurred maps. Both methods can position a document close to similar documents in terms of lexical content, even when the content is expressed with synonyms rather than the same words. CA is slightly better in this respect because visualization of proximities is unclear in LSA due to cosine distance. CA has other advantages, such as numerous numerical and visual interpretation aids, particularly confidence ellipses. Furthermore, CA analyzes data in reference to the independence model, which opens the possibility of using other models. Additionally, a wide range of CA-related methods are available. These methods allow us to handle more complex textual data, such as multiple textual data tables (see Chapter 6) and mixed textual-contextual data tables (see Chapter 7). These methods provide new perspectives, especially when analyzing open-ended responses in different languages and studying the relationship between responses and respondent characteristics (see Chapter 9).

One advantage of CA is its visualization capabilities. These visualizations convert chi-square distance to classical Euclidean distance, which is the only type perceivable by the human eye. In the case of large corpora, both in terms of the number of documents and words, the shape of the clouds is informative and relevant. CA's projection onto a low-dimensional space reveals intrinsic patterns in the data through visual inspection, which is an advantage over LSA. Furthermore, in CA, a word or document is surrounded by its most closely related words or documents. As seen above, CA can position a document close to similar ones even if they are not expressed with the same words, but rather with synonyms.

3

Clustering methods

3.1 Introduction

Cluster analysis and classification are both techniques used in machine learning and data analysis, but they have different purposes and characteristics.

Cluster analysis, also known as *clustering*, is a set of unsupervised classification techniques designed to identify inherent patterns, similarities, or relationships within the data. The objective is to identify hidden patterns in data without relying on pre-labeled output categories. To accomplish this, similar objects will be grouped together in the same cluster, while they must clearly differ from objects in other groups. Huygens' principle variance provides insights to understand the source of variability within a dataset. Total variance is decomposed into two components: between-group variance (inter-group or explained variance) and within-group variance (intra-group or unexplained variance).

This decomposition is essential for evaluating the effectiveness of clustering algorithms because the structure of the groups is not known, and often not even the number of groups. This approach ensures that the classes are generated dynamically, making them more flexible to significant changes in the data. In principle, as an unsupervised technique, the performance of the model will not be evaluated using a set of tests to compare with an unknown true structure, although such tests are often employed for the purpose of selecting the appropriate number of groups.

When data is unlabeled, cluster analysis can be used to group similar elements, providing a comprehensive understanding of the underlying patterns and relationships present in the dataset. These objectives are achieved through two key processes: the synthesis of information and the reduction of high dimensionality. The former simplifies the modeling process, while the latter improves the interpretability of the results.

Once the distance between points and agglomeration method has been established, minimal user intervention is required during the clustering process. The number of clusters and subsequent data labeling represent the only user interventions to consider. Although the labeling process can be supported by an automatic description of the clusters obtained from internal or external variables, it is typically conducted by a human expert. This process is referred to as *consultative learning*.

DOI: 10.1201/9781003397472-3

On the contrary, *classification* is a supervised learning technique that aims to build a model based on previous information known to the researcher, predefined labels or categories, or sometimes from the classes discovered from data patterns. Once the classification criteria have been established, the problem can be solved using a loss function. The goodness of fit of the model can be easily calculated from the analysis of the confusion matrix, establishing relationships between correct classifications as true, correct classifications as false, false positives, and false negatives. In certain instances, the implementation of supervised learning methods with a substantial amount of unlabeled data can be costly or time-consuming. In such scenarios, semi-supervised learning methods can serve as a viable solution. During the training data collection process, a small amount of data is manually labeled by humans, while the rest of the unused data is automatically labeled (Nigam et al., 2000; Blum and Mitchell, 1998).

3.2 Clustering textual data

The classes obtained through a cluster analysis can be considered a reflection of the underlying structure of the entities that represent the data.

Among the input data types, we highlight the Document × Term (DTM) matrix and the correspondence analysis (CA) factor coordinates obtained in Chapter 2.

From a statistical perspective, DTM crossing documents (cases) and words (variables) exhibits distinctive characteristics that differentiate the selected cluster methodology employed from those used in non-textual analyses. In a DTM, documents tend to be numerous, except when crossing categories and words in an *aggregated lexical table* (ALT). In contrast to many non-textual analyses, the variables (words or segments) are not defined *a priori*, but rather derived from the results of the analyzed texts. It is to be expected that different texts will provide different variables.

DTM is often a sparse matrix (except for the ALT case), with low frequencies for words in documents, especially when using open-ended questions in questionnaires. A cluster analysis of the rows (documents) or columns (words) can be performed on the DTM data. In both cases, the high dimensionality implies a high degree of complexity in the interpretation of the results, and many algorithms are not effective with high-dimensional data.

The process of analyzing the shape of a cloud, whether it be documents or words, also entails identifying unusual and noteworthy individuals, or outliers, which are positioned significantly away from the cloud's center of gravity (CoG). However, the text data field reveals a substantial number of clusters compounds only for a few elements with unclear labels. These are not easily classified using predictive models. In some cases, an individual or document may possess its own vocabulary or characteristics that distinguish it from others. One option for mitigating the effects of outliers is to control the elements of the DTM matrix that may introduce instability, both in the CA and in the cluster analysis, as well as in the documents and words.

In such scenarios, it would be advisable to establish a minimum number of documents in which a word must appear in order to be selected during the construction of the DTM matrix. These elements can be removed from the analysis or be considered only as supplementary, without intervening in the construction of the model and subsequently being incorporated as supplementary elements.

The TextData function of the **Xplortext** package (Bécue-Bertaut et al., 2024) allows users to set a minimum frequency threshold (Fmin) for a word to be included in the model. It should be noted that words with low frequency will not be able to characterize individuals. Similarly, establishing a maximum threshold (Fmax) could be useful. Beyond this limit, the word is considered so common in all documents that it will not be useful for characterization. With the exception of stylometry studies, which are defined as the application of linguistic stylistic analysis, a considerable number of words in the texts do not provide significant meaning by themselves and, as a result, serve to increase the dimensionality of the study. These stopwords can be common stopwords (stop.word.tm), such as prepositions, articles, pronouns, and conjunctions, or common words in the specific context analyzed that can be defined by the user (stop.word.user), for example, *think* when an opinion is required. Numbers can also be removed (remov.number). Obviously, these words depend on the language in which they are expressed. This procedure offers the added benefit of allowing for a substantial reduction in the number of columns of the DTM matrix.

It should be noted that the analysis can be performed using the original or the lemmatized texts. When the input data for cluster analysis comes from CA, we will have a words × dimensions matrix and another documents × dimensions matrix.

A co-occurrence matrix is a data structure that records the frequency of occurrence of a given row modality in conjunction with a specific column modality (see Chapter 1). In this context, the DTM matrix is an example of such a matrix when the windows that define the maximum distance to establish a relation between the two words is equal to the length of the document. If the objective is to analyze the nature of the relationships between rows (documents), the co-occurrence table (documents × documents) will represent the number of words that appear simultaneously in each pair of documents.

The words × words matrix indicates the number of documents in which the specific pair of words appears at least once together. A co-occurrence matrix can be constructed for each document or word, and the total co-occurrence matrix can be obtained by adding the partial matrices (Cariou and Qannari, 2018). It is important to note that this matrix is particularly meaningful for interpreting open-ended questions in questionnaires, as their responses often result in sparse data (i.e., many zero or low frequencies). However, when the documents are very large, the co-occurrence matrix will contain many combinations of words with a frequency equal to the number of documents. This may result in interpretability issues. In any case, both matrices can be obtained by multiplying the DTM matrix by the transpose matrix and vice versa.

If the objective is to reduce the dimensionality of the data set through SVD techniques (see Chapter 2), it is preferable to start from the DTM matrix to interpret the relationships between words and documents. These relationships are lost when analyzing symmetric matrices containing co-occurrences. The DTM contingency table can be decomposed using SVD methods. The cluster analysis of the CA coordinates associated with the first factors (Lebart and Mirkin, 1993) allows for the control of the random part, excluding the last factors.

The elimination of non-significant factors has the effect of reducing the importance of the outliers and, consequently, their number in the cluster analysis results. It is evident that the selection of the number of factors can influence the outcomes. In order to ensure the consistency and reliability of the results, it is advisable to identify a threshold beyond which these results remain stable. It is important to note that the CA coordinates of the factors are centered, but that the dispersion of each factor (inertia) is smaller as the number of factors analyzed increases (the first factor is the most important). Therefore, they should not be typified (scaled) to ensure that the importance of each factor is preserved during the performance of cluster analysis.

As outlined in the Chapter 1, the GloVe model can be another source of data providing a co-occurrence matrix. It was also noted that Word2Vec offers word-level embeddings with non-orthogonal dimensions as input data. Cluster analysis for words can be performed directly on these matrices, while cluster analysis on documents will require some transformations.

In the absence of any *a priori* information regarding the shape of the clusters, a minimum knowledge of the domain is necessary to determine some of the input parameters, such as the minimum threshold (Fmin), the number of factors to be retained from the CA, and the criterion that allows the choice between different numbers of clusters.

3.2.1 Classification of clustering methods

As von Luxburg et al. (2012) point out, the main challenge in choosing a clustering technique is the difficulty of evaluating a clustering algorithm without considering the context of its intended use.

It is also essential to have a clear understanding of the properties of the selected clustering method to ensure an accurate interpretation of the structure it reveals. They discuss the existence of a taxonomy of clustering problems and that there is not one right clustering solution or best clustering method. Nevertheless, the inability to employ external cluster validation due to the lack of labeled classes and the uncertainty regarding the *true* number of clusters in advance does not preclude the use of internal validation methods to select the number of clusters and an appropriate clustering algorithm. Additionally, relative cluster validation enables the assessment of other parameters, allowing for the comparison of results within a given context.

Furthermore, the presence of individuals belonging to one group or another can be understood as a degree of belonging through shared characteristics (fuzzy), rather than seeking a perfect classification when considering mutually exclusive clusters. In disciplines such as ecology, the identification of discontinuities entails the delineation of groups of similar objects, occasionally disregarding the potential presence of intermediate individuals (subspecies) between these groups.

The types of distances that can be used, as well as the aggregation criteria in cluster analysis, are determined by the types of input data. The most common distances employed for qualitative variables are the chi-square and cosine. For quantitative variables: Euclidean, Manhattan, and Mahalanobis. In cluster analysis, whether supervised or unsupervised, the methods usually assume a specific type of data, as well as the aggregation criteria of individuals and clusters. This means that each clustering method can only be applied to one type of data.

In this chapter, we will consider that the input data when performing a cluster analysis comes from numerical variables in order to be able to show comparable results.

Finally, some cluster analysis can consider whether to assign equal weight to each document or word, or whether to assign a weight based on document length or word frequency.

This chapter focuses on problem-solving with clustering in textual data analysis and the use of clustering methods related to the objectives. A detailed review of clustering techniques and algorithms can be found in several works (Xu and Wunsch, 2005; Berkhin, 2006; Jain, 2010; Hennig et al., 2015).

Clustering methods are general approaches or frameworks for grouping data into clusters, while cluster algorithms are the procedures used to perform the clustering methods. A variety of classification systems can be employed to organize methods, with distinctions made according to the nature of the data, the algorithms utilized, and fundamentally, the objectives pursued. Henceforth, the individuals in the cluster analysis, whether documents or words, shall be referred to as *points*.

The initial categorization of the *exclusive*, *hard clustering*, or *crisp* methods is intended to classify each point within a single cluster. In this scenario, it is not possible for a point to belong to two distinct clusters simultaneously. The majority of clustering methods can be categorized within this category.

In contrast, *soft clustering* or *fuzzy clustering* methods assign probabilities of membership of each point to each cluster. This approach facilitates the creation of more realistic clusters, and in textual data analysis, one point-word can be related with several clusters when it shares meaning, as well as when it may appear in different contexts.

The third category of methods is *grid-based* clustering, which involves dividing the data space into a grid structure and grouping cells based on the density.

One of the most widely used classifications is as follows:

- Hierarchical. Connectivity-based. Clusters should be constructed based on how close or connected the data points are. It includes two main approaches: agglomerative and divisive.
- Partitioning. Centroid-based. K-means, k-medoids (PAM), and CLARA (Clustering Large Applications).
- Density-based approaches, also known as density estimation, are methods used to identify the way in which the data is distributed within a specified space. This approach is a variant of connectivity-based methods.
- Distribution-based or model-based. The model is regarded as a probability distribution, with the data points deriving from either a Gaussian mixture model (GMM) or an expectation-maximization (EM) process.
- Grid-based methods. STING (Statistical Information Grid), CLIQUE (Clustering in Quest) or WaveCluster.
- Soft clustering. Fuzzy c-means (FCM), model-based, GMM, self-organizing maps (SOM) and soft spectral clustering.

3.2.2 Cluster validation

One of the most crucial aspects of cluster analysis is the assessment of clustering results to identify the optimal partitioning that aligns with the underlying data structure. However, it should be noted that no optimal clustering exists. Different methods, algorithms, or even different configurations of the same algorithm can produce different partitions. In addition, a significant proportion of clustering algorithms require that the number of clusters be specified in advance.

The selection of the number of dimensions to be included as input data for a cluster analysis may have a significant impact on the resulting outcome and the validation. In SVD methods, the first factors collect most of the data structure. As less significant factors are included, the random component, or noise, increases, resulting in more outliers and smaller clusters. This reduction does not apply to other types of data, such as CBoW or GloVe, due to the nature of their dimensions.

From the perspective of cluster interpretation, particularly in the context of document analysis, the ability to understand the data structure is of paramount importance. This ability enables experts within a specific discipline to identify and delineate a more substantial number of clusters in comparison to those identified by non-experts.

In this context, cluster validation techniques are typically classified into three groups (internal, relative, and external validation), although the distinction is not always clear.

3.2.2.1 Internal validation

The process of comparing different sets of clusters in the absence of external knowledge is referred to as an *internal quality* measure. The objective may be to ascertain the number of clusters or the clustering method. However, it is important to note that there is no optimal solution, because the result is strongly dependent on the specific data set under consideration (shape of the points). In our case, the number of factors selected from the CA results is considered here in order to analyze the effect on the selection of the number of clusters.

Milligan & Cooper (1985) proposed 30 different *Cluster Validity Indices* (CVI) to find a stopping rule that will allow for the selection of the number of clusters in a hierarchical clustering algorithm (HCA). Other studies have also identified different numbers of clusters using various CVIs, suggesting that no single CVI consistently outperforms the others (Arbelaitz et al., 2013). However, the implementation of this procedure is not always straightforward due to the monotonic or degenerative trend observed in some CVI.

Many of the CVIs are based on the decomposition of Huygens' *total sum of squares* (TSS) into the *sum of squares within the group* (WSS) and the *sum of squares between groups* (BSS). The objective is to divide the set of documents so that, within each cluster, the documents use similar vocabulary, and from one cluster to another, the vocabulary is relatively different. This implies that the goal could be to maximize *BSS*, which is equivalent to minimizing *WSS*.

The total scatter matrix $\mathbf{T}$ quantifies the dispersion around the barycenter $\mathbf{b}$ of the data matrix, $\mathbf{X}$. Where $\mathbf{W}$ is the diagonal matrix of weights corresponding to the row marginals of $\mathbf{X}$.

$$\mathbf{T}_{p \times p} = (\mathbf{X} - \mathbf{b})^{\mathbf{T}} \mathbf{W} (\mathbf{X} - \mathbf{b}).$$

The input data used in cluster analysis must be scaled if the variables are expressed in different units. If the factor coordinates of a CA are used, they are already centered and scaling is unnecessary. Scaling would cause the factors to appear equally important (i.e., have the same inertia), and thus weight importance (represented by their eigenvalues) would be lost.

This way, the $\mathbf{T}$ matrix can be simplified as follows:

$$\mathbf{T}_{p \times p} = \mathbf{X}^{\mathbf{T}} \mathbf{W} \mathbf{X}.$$

The trace of $\mathbf{T}$ is TSS expressed as the distance of each observation x_{ij} to the barycenter of the variable $\bar{x}_j$ with different weightings $f_{i.}$ for each point:

$$tr(\mathbf{T}) = TSS = \sum_{i=1}^{N}\sum_{j=1}^{p} f_{i.}(x_{ij} - \bar{x}_j)^2.$$

Additionally, since the factorial axes are orthogonal, the total inertia (TSS) is simply the sum of the inertias of each individual axis. Therefore, TSS can be computed directly from the (document or word) coordinates F_{is}, as these coordinates are already centered (i.e., $\bar{F}_s = 0$), where S is the number of factors. Let $\mathbf{F}$ be the centered coordinate matrix (derived from $\mathbf{X}$). The TSS can then be partitioned into K groups, where I_k denotes the number of points in cluster k:

$$TSS = \sum_{i=1}^{N}\sum_{s=1}^{S} f_{i.}F_{is}^2 = \sum_{k=1}^{K}\sum_{i=1}^{I_k}\sum_{s=1}^{S} f_{i.}F_{is}^2.$$

Within-group sum of squares WSS can be computed as the weighted squared distances from the point to the center of gravity of the cluster:

$$WSS = \sum_{k=1}^{K}\sum_{i=1}^{I_k}\sum_{s=1}^{S} f_{i.}(F_{is} - CoG_{ks})^2.$$

Between clusters sum of squares BSS is the sum of the weighted squared distances from each cluster's center of gravity to the overall center of gravity of the data:

$$BSS = \sum_{k=1}^{K}\sum_{s=1}^{S} f_k(CoG_{ks})^2.$$

$Trace\_W \equiv WSS$ is one of the most widely used CVIs. Its value typically declines in a monotonic fashion as the number of clusters increases, so the selection of this number is typically made on the basis of the largest difference between two contiguous indices. WSS is likely to have good performance if all clusters have the same dispersion.

Ball-Hall BH index is the average of the WSS, providing a similar interpretation:

$$BH = \frac{1}{K}\sum_{k=1}^{K}\left[\left(\sum_{i=1}^{I_k}\sum_{s=1}^{S} f_{i.}(F_{is} - CoG_{ks})^2\right)/w_k\right].$$

Where w_k (the total weight of cluster k) is defined as:

$$w_k = \sum_{i=1}^{I_k} f_{i.}.$$

Banfield-Raftery BR is more appropiate than WSS and BH when the clusters are hyperspherical but of different sizes:

$$BR = \sum_{k=1}^{K} w_k log(\sum_{i=1}^{I_k} \sum_{s=1}^{S} f_{i.}(F_{is} - CoG_{ks})^2 / w_k)).$$

Calinski-Harabasz CH index also known as the Variance Ratio Criterion VRC or pseudo ANOVA F test, searches to maximize the ratio between BSS (maximize) and WSS (minimize) normalized by their number of degrees of freedom. The value of K that gives the maximum CH index is chosen as the optimal number of clusters.

$$CH = \frac{BSS/(K-1)}{WSS/(N-K)}.$$

Two other indices constructed with WSS and BSS are $LSSR$ Hartigan:

$$LSSR = log\,\frac{BSS}{WSS}.$$

And R-Squared, interpreted as the percentage of inertia explained by the clusters.

$$R^2 = \frac{BSS}{TSS} = \frac{BSS}{BSS + WSS}.$$

R-Squared can be calculated for each cluster considering $TSSk$ and $WSSk$ computed with only the points belonging to cluster k. In this way, it is possible to choose the number of clusters so that there are no clusters with a low explained value.

$$R_k^2 = \frac{BSS_k}{TSS_k} = \frac{BSS_k}{BSS_k + WSS_k}.$$

The number of clusters can be calculated based on several criteria, sometimes not mutually exclusive, such as compactness, separation, and connectivity.

- Compactness, internal cohesiveness or cluster cohesion. CVIs measure how close objects are within the same cluster (homogeneity), minimizing WSS. They can be used when clusters are expected to have dense structures.
- Separation. They measure how much a cluster differs from other clusters (heterogeneity) using cluster centers or pairwise comparisons between points from different clusters. The goal is to maximize these BSS distances.
- Connectivity. The objective is to evaluate a specific point within a designated cluster, considering the spatial location of the other points.

Matrix algebraic operations using the traces and determinants of matrices can be performed to select the number of clusters. Let $\mathbf{WG}_k$ be the weighted within-group scatter matrix, representing the dispersion of the k-th cluster objects around their centroid CoG_k. The total within-cluster sum of squares and cross-products matrix, $\mathbf{WG}$, can be calculated by adding the individual $\mathbf{WG}_k$ matrices. The Friedman-Rubin DR index seeks to maximize this ratio:

$$DR = \frac{|\mathbf{T}|}{|\mathbf{WG}|}.$$

Considering that the total scatter matrix $\mathbf{T}$ is independent of the number of clusters, the objective is mathematically equivalent to minimizing the determinant of the within-cluster matrix, $|\mathbf{WG}|$.

It is also possible to identify a family of indices based on intra- and inter-cluster pairwise distances. Pairwise distance-based CVIs seek to describe cluster compactness (for objects belonging to the same cluster) and inter-cluster separation (for objects belonging to different clusters). Many CVIs are based on Dunn's index. A common variant, using the k centroids $\{CoG_1, CoG_2, ..., CoG_k\}$ and a distance function d, is defined as follows:

$$D_k = \frac{min_{C_r \neq C_s} d(CoG_r, CoG_s)}{max_{k \in K}(diam_k)}.$$

Here, the denominator $max_{k \in K}(diam_k)$ is the maximum diameter of any cluster in the partition. The numerator quantifies separation (e.g., the minimum distance between cluster centroids), and the denominator measures compactness.

Rousseeuw (1987) introduced the silhouette, which evaluates how well each point fits within its assigned cluster by comparing its compactness (closeness to other points in its cluster) and its separation (distance from points in the nearest neighboring cluster). The silhouette measure has three main functions:

- It evaluates the coherence of each individual point's assignment to a given group.
- It can be averaged to assess the overall quality of each cluster.
- The average of all values provides a global measure (the average silhouette width) for the entire clustering.

The silhouette value ranges from -1 to $+1$.

The score for each individual point is interpreted as follows: a high positive value (near $+1$) indicates the point is well-matched to its own cluster and far from other clusters; a value approaching 0 suggests the point lies on the border between two clusters; and a negative value indicates the point may be assigned to the wrong cluster.

These individual scores are often averaged to assess the quality of the clustering, providing either a global average (for the entire dataset) or a cluster average (for all points within a single cluster). These averages are commonly interpreted using thresholds. For instance, many authors consider an average score above 0.7 to indicate a strong structure, above 0.5 a reasonable structure, and below 0.25 a weak or possibly artificial structure.

If the total number of points is not too large, these individual values can also be visualized in a silhouette plot.

The original formulation of silhouette CVI assumes a uniform weighting. In our case, and if we use the coordinates from a CA, we will apply the $f_{i.}$ weighting used above and the functions wcClusterQuality and wcSilhouetteObs of the **WeightedCluster** package (Studer, 2013). Let a_i be the average weighted distance of a point i with the other members of its group k:

$$a_i = \frac{\sum_{j \in k, j \neq i} f_{j.} d_{ij}}{\sum_{j \in k, j \neq i} f_{j.}}; \qquad f_{k.} = \sum_{j \in k} f_{j.}.$$

Let b_i be the minimum weighted distance of a point i to the points in l group other than k;

$$b_i = \min_{l \neq k} \left(\frac{1}{f_{l.}} \sum_{j \in l} f_{j.}\, d_{ij} \right) \quad ; \quad f_{l.} = \sum_{j \in l} f_{j.}.$$

The silhouette of the point i is then computed as follows:

$$s_i = \frac{b_i - a_i}{max(a_i, b_i)}.$$

3.2.2.2 Relative validation

The objective of the relative cluster validation criterion (Vendramin et al., 2010) is to evaluate the clustering structure by varying different parameter values, such as the number of input data factors from a CA or the number of clusters without *a priori* knowledge.

The elbow method is a heuristic technique used in cluster analysis to determine an appropriate number of clusters using a two-dimensional plot. This plot illustrates either the WSS or the percentage of explained inertia, for each number of clusters considered. Intuitively, a higher number of clusters can be expected to imply a monotonic relationship that explains a higher percentage of the total inertia and reduced WSS.

The number of selected clusters will be indicated by the point at which increasing the number of clusters does not significantly increase the explained inertia. One potential drawback of this method is that the choice is subjective. When interpreting the graph, it is important to consider that WSS for random data (i.e., data without group structure) can still produce a curve that looks like an elbow, as it often follows an approximately 1/K shape. This artificial elbow can lead to mistakenly identifying a structure that isn't there.

Therefore, the elbow method is often interpreted in conjunction with other indices, such as the silhouette score and Dunn's index, to make a more robust decision. Additionally, the presence of outliers can also be problematic, as they can distort the WSS plot and mask a true elbow. The k-means algorithm is sensitive to random initialization, which can result in finding a poor local minimum, especially in the presence of outliers. In some cases, the presence of outliers can result in a solution with $K+1$ clusters that explains less inertia than the solution with K clusters.

Figure 3.1 shows the average silhouette obtained for the *Health* database varying the number of clusters (from two to seven) and the number of factors (from two to 20) and applying an HCA.

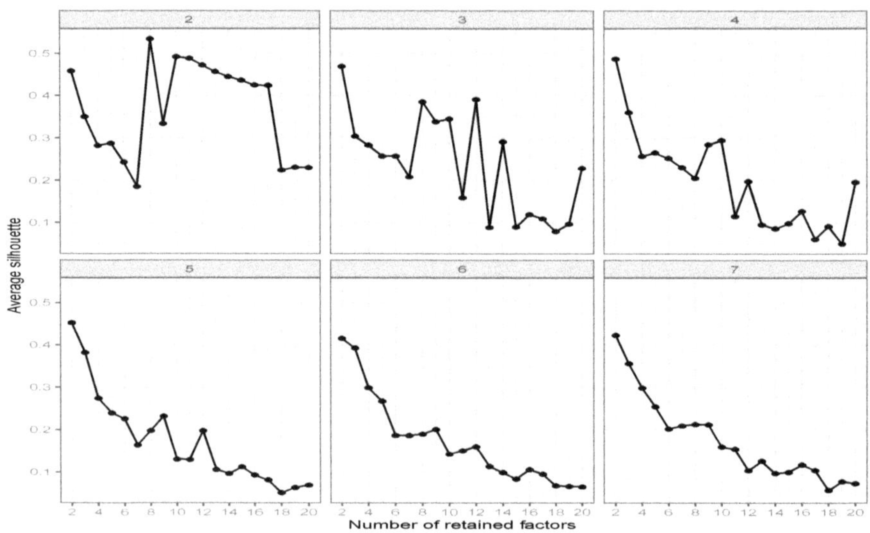

Figure 3.1: Average silhouette by clusters (from two to seven in grey) and retained factors.

For a small number of clusters (two and three), there is a decrease in the percentage of variance explained as the number of factors increases until seven factors, from which there is some increase for two clusters and certain stability for three clusters. This may be due to a factor that is highly correlated with a small number of documents. After three clusters, the decrease in the average silhouette tends to be monotonic by incorporating factors with a greater random component that reduce the structure of the model obtained only with the first factors.

As illustrated in Figure 3.2, the relationship between the number of retained factors and the percentage of explained inertia is inverse from two to seven clusters. It is evident that increasing the number of clusters has a more significant impact as the number of retained factors decreases.

Figure 3.3 shows the percentage of points (documents) with very bad representation in the cluster (with a negative value of silhouette). If only a few clusters are chosen, the percentage of documents that are far from the

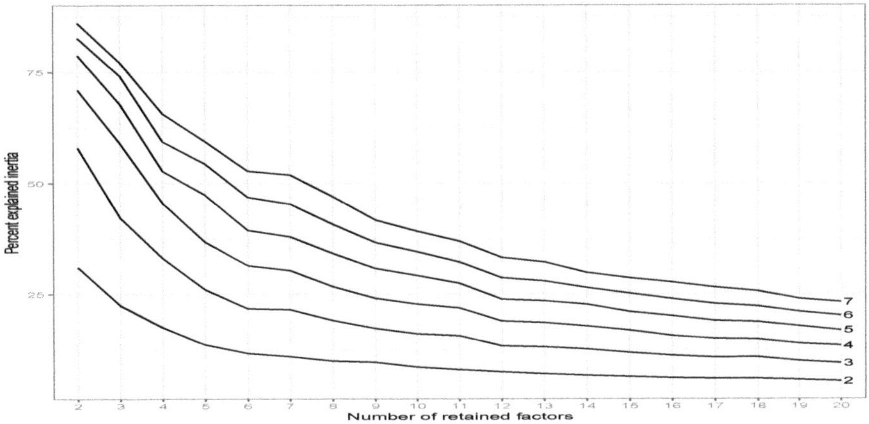

Figure 3.2: Percentage of explained inertia by clusters (from two to seven) and retained factors.

cluster they belong to does not increase significantly as the number of factors increases (first row in Figure 3.3).

Incorporating more factors leads to greater randomness and a loss of structure. This is indicated by the increase in the percentage of documents with negative silhouette values for the same number of clusters (row two of the Figure 3.3). It is evident that the percentage of documents not well represented by the cluster to which they belong increases for two clusters when selecting five and six factors, obtaining few non-negative and stable values as the number of factors increases over six. There are some significant jumps in the choice of three and four clusters, depending on the number of factors chosen. In the second row of this figure, starting from five or eight clusters, there is an increase in the number of misrepresented documents as the number of factors used as input data in the cluster analysis increases.

The objective is to achieve a balance between the average silhouette, the percentage of variance explained, and the poor representation of the documents.

3.2.2.3 External validation

The objective of external comparison indices is to assess the similarity between two cluster configurations.

This validation method uses an external, supervised criterion (like a pre-existing classification) to evaluate the results of an unsupervised clustering algorithm.

These indexes facilitate the analysis of the similarity between the results of cluster analysis performed using different methodologies, and the identification of potential reasons for any observed differences. In this particular

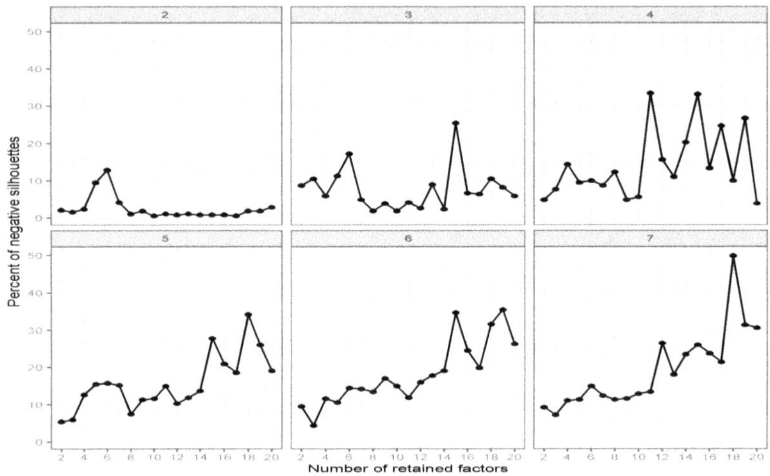

Figure 3.3: Percentage of points with negative silhouette by clusters (from two to seven in grey) and retained factors.

instance, the indexes are characterized by a comparative perspective rather than a validation nature.

If the objective is to compare the results of the classification performed with a different number of factors, these indices can be used to detect jumps and evaluate the appropriateness of the choice.

A first approach is to obtain a confusion matrix. This is achieved by calculating the pairwise relationships between the points of the two clusters to be compared. The **R** library **clusterCrit** (Desgraupes, 2023) can perform 11 external indices.

It is also possible to compare two cluster configurations, even if the number of clusters in each configuration is different, for example, using the adjustedRandIndex (ARI) function (Rand, 1971; Hubert and Arabie, 1985) of the **mclust** R package (Scrucca et al., 2023) and dendrograms using tanglegrams (see Chapter 4 and Chapter 5 for examples).

3.2.2.4 Internal textual characterization validation

The process of clustering documents involves the use of unsupervised classification search to identify underlying patterns and relationships present in the data set. However, it is important to note that this procedure cannot be accomplished in the field of textual data analysis without first labeling the obtained groups. This labeling can be performed on the basis of the socioeconomic data of the respondents in the case of open-ended questions in questionnaires or on the basis of characteristics contained in the metadata of the documents.

Once the number of document clusters has been selected for the chosen methodology, labeling involves determining the words that are characteristic of each cluster through an *internal textual characterization validation* (ITCV). It is reasonable to expect that documents from different clusters will use different terminology. These words can be used to label them.

It is important to note that a given word may appear as a feature in two different groups, usually because it may be used in different contexts. A cluster without characteristic words may represent documents that contain words common to all groups, sometimes corresponding to people who do not have a clear opinion on what was asked in an open question, or that are not specialized documents in the case of texts. In this case, ITCV can consider this solution valid, given that all the clusters are providing information. According to this reasoning, the existence of two clusters without characteristic words does not provide information, since two clusters that in principle are different cannot be labeled with the same label, which is also empty. Therefore, the presence of two clusters with empty labels suggests the presence of an excessive number of groups. A manual review of the characteristic words is necessary to ensure that they do not contain words that are not tools, i.e., that provide information in the labeling process.

Although the interpretation of the ITCV in this section is still subjective, it can assist in determining the number of factors when the input data are the coordinates of a CA and the number of clusters. These indices are independent of the clustering method applied, some of which are listed below.

Similarly, it is not logical to construct groups where the distance of the points from their center is so large that it is impossible to characterize them. The use of individual silhouette values allows for the detection of documents or words that behave as outliers.

3.3 Example. Corpus issued from an open-ended questionnaire

In 1989-1990, the Valencian Institute of Public Health (IVESP) surveyed 513 people aged 14 and over about their attitudes and opinions related to health and lifestyle. The questionnaire asked people to answer the question *What does health mean to you?* in their own words. The main objective in this chapter is to discover and describe the possible existence of individuals who share the same vision of the concept of health.

The responses in Spanish have been translated into English. In the pre-treatment phase, the capitalized words after a period were transformed into lowercase, the contractions were recoded to their original words and subsequently lemmatized with the **udpipe R** package.

Finally, a DTM with 394 non-empty answers in rows, 804 different words in columns was constructed, yielding a total of 10,303 occurrences. The mean length for the non-empty answers was 26.15 words.

3.3.1 Number of factors from CA

The first decision to be made is the choice of the number of CA factors to be used as input data for the cluster analysis. The dimRestrict function in the **FactoInvestigate** package is designed to calculate the number of factors that are significant. It considers factors to be significant as long as their eigenvalues are above a threshold determined by randomness.

This idea is applied to analyze the effects of setting a minimum frequency for a word to be included in a study. Figure 3.4 shows how the number of factors goes down as the minimum threshold goes up.

It also demonstrates that, starting at a minimum frequency of eight, the results vary when stopwords are included in the analysis and when they are removed (script **chap3.a**).

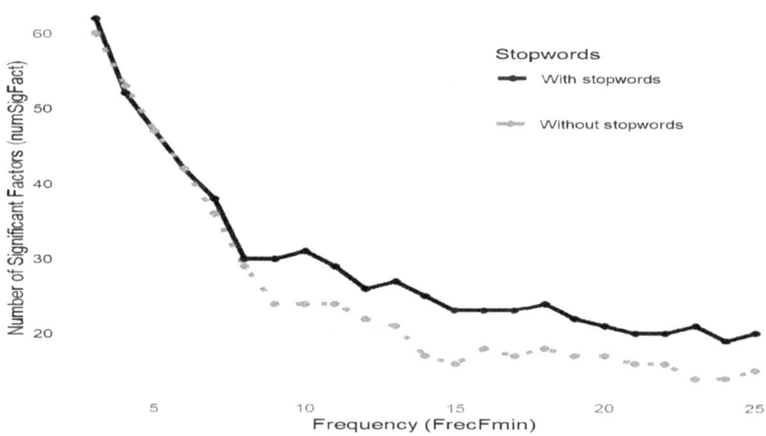

Figure 3.4: Proposal of factors using FactoInvestigate package.

3.3.2 Clustering documents

The process of clustering documents involves the analysis of each document as a unit. The first step was discussed in Chapter 1, which outlined as one of its objectives the extraction of meaningful words from the set of documents.

Once a corpus has been preprocessed (tokenization, lowercasing, stopwords removal, stemming/lemmatization, and threshold selection), the DTM matrix is obtained using a Bag-of-Words (BoW) approach. Each element of the DTM matrix indicates how many times a word appears in a document, allowing us to build as many vectors as there are documents. The columns represent the selected words from the vocabulary. However, as previously mentioned, the DTM matrix does not allow for the order of words to be reconstructed during an analysis. For this reason, frequent criticism refers to a loss of syntactic and semantic understanding when processing documents. In some cases, the same word can have different meanings in different sentences or corpora. This ambiguity is less prevalent in open-ended questions and specialized texts with narrow contexts. The problem becomes more significant when the texts belong to broader contexts.

As previously mentioned in Chapter 2, documents can be evaluated based on their lexical content. This content can be measured using the characteristic vocabulary of each document in the DTM matrix. The objective is to represent the proximity between documents, between words, as well as the relationships between documents and words in reduced dimensional spaces using SVD methods as latent semantic analysis (LSA) or CA. In these spaces, the degree of proximity among documents is positively correlated with the degree of lexical overlap among their contents. The lexical profile of a document is similar within the same cluster and different from another cluster.

A particular instance of cluster analysis is that of ALT, where the number of individuals is reduced to the number of different categories of the aggregation variable.

Once the documents are divided into groups, each cluster can be described using the set of words that characterizes each one. If there are a very small number of clusters, there may be a very large number of words that describe each group. Conversely, if the number of clusters is excessive, there may be clusters that do not have characteristic words. A group of documents can be described by the absence of distinctive words. However, it would not be logical for more than one of them not to have keywords.

This characterization can also be performed using quantitative or qualitative variables available for the documents. In this case, the objective is to use these variables to understand the reasons behind the formation of the clusters, rather than to identify the words that characterize each variable.

3.3.3 Clustering words

Word cluster analysis on the DTM matrix can be performed using the same approach as cluster analysis with documents. In this section, we will consider the specific approach of unsupervised learning algorithms using pre-trained models. As outlined in Chapter 1, there are various processes for vectorization, in which words are converted into a set of numbers (vectors) that capture their semantic meaning.

The central concept posits that the significance of a word is derived from its contextual relationship with other words in its immediate vicinity. The implementation of cluster analysis is performed on these vectors, not on the words themselves.

BERT (Bidirectional Encoder Representations from Transformers) is a pre-trained language model that uses word embeddings developed by Devlin et al. (2019). It employs bidirectional context. The overall process involves several steps (script **chap3.b**).

1. BERT is written in Python language. Python can be installed on your computer as an additional language option or using the **R** function install_miniconda. Additionally, the installation of other necessary packages is required (**r-reticulate**, **transformers** and **torch**).
2. Load a pre-trained tokenizer model to translate our text into a numerical format for processing with BERT.
3. Load our RData in a dataframe and remove NA and empty documents.
4. Create a list of word embeddings for every word in every document. In our case, it is a tibble with 11,961 row-words and 769 columns.
5. Stopwords are eliminated, and words that meet the frequency threshold (set at five in this instance) are retained.
6. Convert the results into a matrix to use clustering techniques. In our case, there are 90 words in rows and 768 dimensions in columns.
7. Determine the optimal number of clusters by employing one of the previously outlined methods. In our example, nine.
8. The desired clustering method, in this case k-means, is applied.
9. Extract the words belonging to each cluster.
10. Words can also be represented in 2D by reducing the high number of dimensions, for instance using t-SNE or UMAP.

Table 3.1 illustrates the allocation of words across the nine clusters. The first cluster encompasses general terms, while the second contains words related to food, nutrition, and exercise. The third cluster consists of words related to mental and psychological problems. The fourth cluster displays positive words associated with happiness. The fifth pertains to personal relationships. The sixth group focuses on taking care of the family, having enough time, a job, and money. The seventh group is described using words related to being fit, healthy, and wellbeing. The eighth group consists of negative words such as bad, ill and suffer. The ninth group is made up of words that have been seen in other analyses, such as alcohol, drink, smoke and tobacco.

3.4 Hierarchical clustering

Hierarchical clustering analysis (HCA) is an unsupervised machine learning algorithm that estimates the degree of relationship between each data point

Table 3.1: Words in nine clusters using BERT.

Cluster	Words
Cluster 1	aspects, define, difficult, essential, influences, level, main, means, takes, taking, terms
Cluster 2	eat, eating, exercise, food, live, living, nutrition, perform, sport
Cluster 3	mental, mentally, physical, physically, psychic, psychically
Cluster 4	calm, cheerful, comfortable, eager, enjoy, feel, feels, happiness, happy, joy, mood, strong, zest
Cluster 5	blood, body, human, mind, oneself, people, person, personal
Cluster 6	bit, care, children, day, family, home, job, life, lot, moment, money, time, world
Cluster 7	condition, conditions, environment, fit, health, healthy, hygiene, illusion, natural, normal, perfect, properly, wellbeing
Cluster 8	bad, depression, discomfort, disease, doctor, hurts, ill, illness, illnesses, pain, sick, stress, suffer
Cluster 9	alcohol, drink, smoke, tobacco

(document or word) and the others, without predefining the number of clusters. It generates a visual tree structure, known as a *dendrogram* or hierarchical tree, which can be used to facilitate rapid and straightforward analysis of how clusters are formed in an iterative data analysis.

In order to measure the relationship between points, it is necessary to define a measure of similarity or a measure of distance. The most common measures are the Euclidean and Manhattan distances. When the coordinates of the result of a CA are used, a quantitative distance such as the Euclidean distance can be applied, even though the original DTM table was a contingency table.

In the context of cluster analysis, the aggregation or divisive method is defined as the process employed to determine the circumstances under which clusters should be merged (AHC or AGglomerative NESting AGNES) or divided (DIVisive ANAlysis DIANA).

For your reference, the script of this section is located in the **chap3.c** file.

3.4.1 Agglomerative hierarchical clustering (AHC, AGNES)

AGNES is one of the most used unsupervised learning techniques to group similar data points into clusters by building a hierarchy (a tree-like structure). The approach is characterized by a bottom-up strategy, wherein the initial step involves the establishment of clusters corresponding to the total number of points, which can be either documents or words. Each point is designated as its own cluster.

The construction of a dissimilarity matrix involves the computation of the pairwise distance (frequently the Euclidean distance) between all points. This approach is subject to a key limitation: it requires substantial memory allocation, particularly when the number of points is high. The subsequent step involves joining the two closest points to establish a cluster comprising two elements. The distance matrix is recalculated.

A critical decision that must be made when performing a cluster analysis is how to select the aggregation method. Lance and Williams establish the relationship between seven aggregation strategies using a linear formula with four parameters (Lance and Williams, 1967):

- Single-linkage or Nearest neighbor method. The dissimilarity between two clusters is calculated as the shortest distance between any single point in the first cluster and any single point in the second cluster. The two clusters with the shortest distance will be joined, without considering the rest of the points. This method has the advantage of being able to find complex structures that are not spherical or elliptical. However, it can merge clusters that are distant but with some points close to them, as may be the case of outliers, being very sensitive to them. Therefore, it is common to observe a chaining effect that is easily identifiable when the number of dimensions is small.

- Complete-linkage or Furthest neighbor method. As with the single-linkage method, complete-linkage requires the computation of all the pairwise distances. The distance between two clusters is defined as the maximum distance between any single point in the first cluster and any single point in the second cluster. The two clusters with the smallest maximum distance are then joined. It prevents the chain effect, building compact clusters with similar size. However, it is highly sensitive to outliers and can easily disrupt clusters formed by multiple points. The analysis often produces clusters characterized by low intra-group sum of squares, which is indicative of compact clusters.

- Unweighted Pair Group Method with Arithmetic Mean (UPGMA). The distance between two clusters is determined as the arithmetic mean of all pairwise distances between the points in one cluster and the points in the other. This method, implicitly weights by cluster size when calculating the new distance to a third cluster, giving more influence to the larger clusters. It's a good compromise, being more robust to outliers, less affected by chaining and favoring spherical shapes.

- Weighted Pair Group Method with Arithmetic Mean (WPGMA). This method treats both clusters as equally important when merging, regardless of their size. It can provide disproportionate influence to small, early-formed clusters.

- Ward's or Minimum variance method. Ward's method aims to merge the pair of clusters that leads to the smallest loss of information when two different points or clusters are joined. The loss of information is evaluated by the increase in WSS when the number of clusters decreases, measured through the intra-group distances.

In older software versions, the WSS was sometimes calculated using non-squared Euclidean distances, which favored the merging of small clusters. The Ward.D2 method works with squared Euclidean distances, implementing the original paper by Ward (1963), producing clusters of similar size while avoiding the chaining effect. It tends to build spherical and compact clusters, but is sensitive to outliers.

- There are two additional methods that may be considered: the centroid and median methods.

For the purpose of comparing the following analyses, the *Health* database has been utilized, with the restriction that only the coordinates of the first two factors of the CA were considered. In this section, the **Xplortext** package is employed to retrieve the lemma of the text, removing stopwords and establishing a frequency threshold of five in at least three documents. The Ward.D2 aggregation method is applied.

As illustrated in Figure 3.5, the tree-like structure of documents in the first factorial plane is represented as a dendrogram, showing how clusters are merged. Three large groups of documents are observed for a height of approximately 0.1. By reducing the height to 0.05, the number of clusters increases to five, with Cluster 7 consisting of a small number of documents visible on the left.

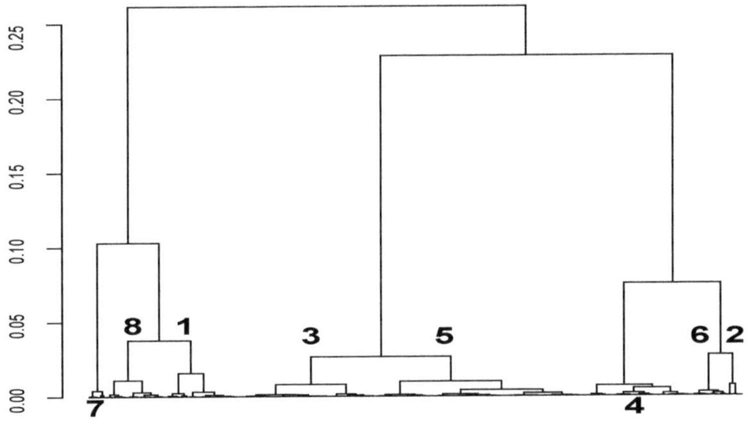

Figure 3.5: AGNES dendrogram.

The elbow chart has been developed to determine the number of clusters that should be considered when computing the percentage of the inertia explained (see Figure 3.6). Other indices described in this Chapter, such as silhouette, can also be used. By selecting five clusters, the percentage of inertia explained is 75.66, increasing to 86.07 in the case of eight clusters.

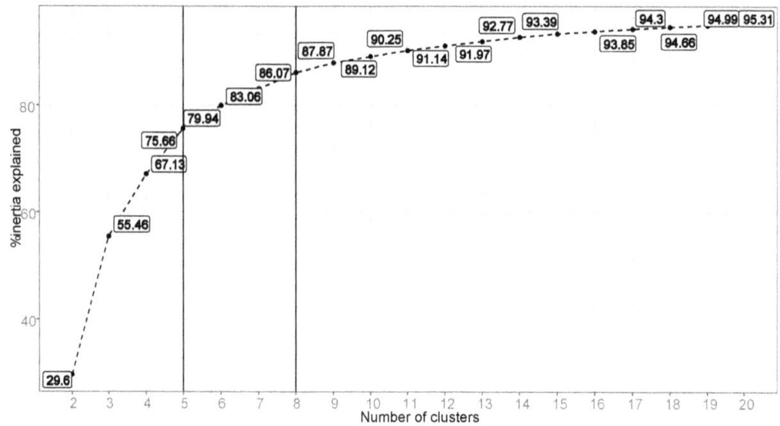

Figure 3.6: AGNES elbow.

In order to evaluate the selection of eight clusters, Table 3.2 presents the results of the intra-cluster analysis of documents using AGNES. Cluster 5 contains 127 documents, with an explained percentage of 83.40%. The smallest clusters are numbers 2 (nine documents) and 7 (eleven documents). All the percentages explained are above 50%, with clusters 3 (54.04%) and 1 (55.71%) being the smallest.

Table 3.2: Intra-cluster results for documents using AGNES with 8 clusters.

Cluster	Docs	Ocurrences	ss.tot	ss.intra	ss.between	%Explained
1	43	419	0.0580	0.0257	0.0323	55.71
2	9	59	0.1324	0.0077	0.1246	94.15
3	81	890	0.0307	0.0141	0.0166	54.04
4	62	714	0.0698	0.0206	0.0492	70.52
5	127	774	0.1322	0.0219	0.1103	83.40
6	21	230	0.1184	0.0065	0.1118	94.49
7	11	41	0.1794	0.0062	0.1733	96.55
8	36	251	0.1616	0.0202	0.1415	87.53

The final phase involves characterizing the obtained clusters, determining their characteristic words, and identifying exogenous variables that explain why the clusters were constructed based on similar lexical profiles. This phase also involves characterizing the axes based on the clusters or extracting the most representative documents for each cluster. This process is especially necessary if the results of the cluster analysis will serve as the basis for subsequent document classification, since uncharacterized clusters will have problems when attempting to be predicted.

As illustrated in the dendrogram (Figure 3.5) and the elbow plot (Figure 3.6), the eight-cluster configuration may be excessive, as several groupings emerge within relatively similar groups, joining at a low height.

If this were the case, some clusters would be expected to share characteristic words due to similar lexical profiles and personal characteristics among the respondents in the survey. The resulting characterization of each cluster is an indicator of the quality of the obtained grouping.

Next, we will consider three distinct groups, analyzing whether they can be effectively characterized and whether they share common elements.

- Clusters 2, 6, and 4 with 92 respondents.

  Cluster 2 appears to represent a dimension of *health and lifestyle behaviors*, encompassing substances (*alcohol, drink, drug, smoke*) and activities (*eat, sport*) that are integral to daily human function. The words *essential* further suggests this cluster relates to baseline human needs and habits. There are only nine respondents, characterized to be men, separated, in good health, under the age of 20 or between the ages of 36 and 50. They assert that they have a high income and are homemakers or unemployed.

  Cluster 6 contains 21 people who share many words with those in cluster 2, including *alcohol, drink, drug, smoke, eat, sport*, and adding *tobacco, exercise*, and *food*. They include the topic *take care oneself*. They are also characterized by being male, under 25 years old, single or separated, with varying degrees of health, and in permanent employment.

  Cluster 4 shares some words with cluster 6, such as *take care oneself, exercise* and *eat*. This cluster, with 62 respondents, and the words they contributed are consistent with expectations for such a group, including terms such as *life, live, normal, natural, much, longer*, and the negative word *stress*. As in clusters 2 and 6, most of the respondents are male, but in this case, they are married. They say they are in good or very good health, have an average income, and between 21 and 35 years old.

- Clusters 3 and 5 have the highest number of respondents, with 81 and 127, respectively.

  Cluster 3 contains generic words such as *can, know, health* and a segment that had already appeared previously *important thing*, three words very related as *job, work*, and *money*. Due to the age of the respondents (over 50), the word *children* may refer to children and grandchildren. The profile in question describes a married or widowed woman who is either unemployed or retired, with a variety of responses regarding the quality of her health.

  Cluster 5 includes positive words such as *main, feel, well, good*, and *cheerful*, but is characterized by a high number of words indicating health problems: *discomfort, eager, hurt, ill, illness, mood, pain, sick*, and *suffer*. They are married or widowed women, over 50 years of age, with average income, pensioners, homemakers, or unemployed, with fair or poor health.

- Cluster 7, 8, and 1.

  Cluster 1 has only 11 respondents. Clusters 8 and 1 have a medium size, with 36 and 43 cases, respectively.

  Cluster 7 is only chacterized by four words: *mental*, *physical*, *state*, and *wellbeing*.

  Cluster 8 has the same four words than Cluster 7, adding *fit condition*, *body*, *perfect*, *balance*, *level*, and *disease*.

  Cluster 1 focuses on people: *person*, *personal*, *problem*, *difficult*, *level*, *condition*, *body*, *mental*, *physical*, and *physically*.

  The three clusters are characterized by being male, single (Cluster 1 also includes widowers), in good or very good health, employed, aged between 21 and 30 (Clusters 1 and 7), and under 20 in Cluster 8. Income is medium-high in Cluster 8 and higher in the other two clusters.

3.4.2 Divisive hierarchical clustering (DIANA)

While the AGNES method employs a *down-top* approach, starting with as many clusters as cases, DIANA method follows a reverse process. It begins with a single group and performs partitions using an iterative process, obtaining one more cluster in each iteration.

The objective is to split the most heterogeneous cluster in each iteration. The result can be represented in a dendrogram. To determine the optimal number of clusters, a methodology similar to AGNES can be employed.

It is possible to find various DIANA algorithms, such as the one described by Reinert (1983), which is used in the Iramuteq package (*Interface de R pour les Analyses Multidimensionnelles de Textes et de Questionnaires*). In this section, the **Xplortext** package is utilized with the same parameters as in the AGNES example.

Figure 3.7 shows the DIANA dendrogram that appears to differ from the one for AGNES. In order to make a comparison, the two results were globally analyzed. The cophenetic correlation coefficient provides a value of 0.495, Baker 0.654, and ARI 0.625, indicating that the relationship between both configurations is considerable, and the clustering results are reasonably consistent across the two methods, considering that they are performed with 390 cases and eight clusters.

The elbow method for DIANA shows a slight increase in the proportion of explained inertia for the first two factors compared to AGNES (from 29.60 to 31.42 for two clusters and 55.46 to 59.15 for three clusters). There are no significant differences in the curves, with the values for DIANA in clusters 4 to 8 being: 66.02, 70.32, 77.43, 84.48, and 86.00.

In order to make a comparison between the DIANA outcomes and those from the eight AGNES groups, the Table 3.3 reveals only minor differences when analyzing the intra-cluster results.

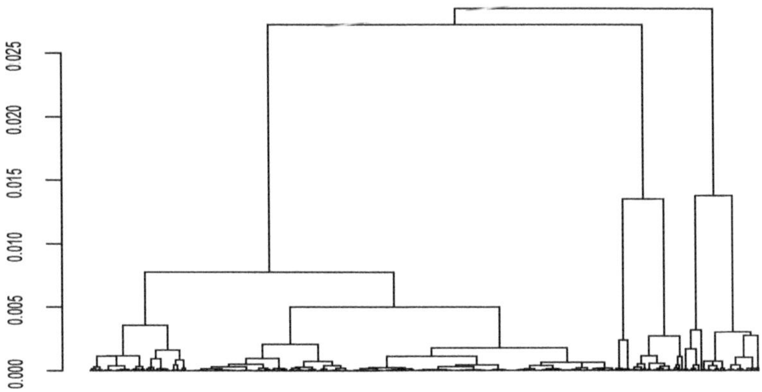

Figure 3.7: DIANA Dendrogram.

Table 3.3: Intra-cluster results for documents using DIANA with 8 clusters.

Cluster	Docs	Ocurrences	ss.tot	ss.intra	ss.between	%Explained
1	32	301	0.0378	0.0084	0.0294	77.84
2	9	59	0.1324	0.0077	0.1246	94.15
3	95	1038	0.0560	0.0262	0.0298	53.18
4	154	1050	0.1445	0.0323	0.1122	77.66
5	24	269	0.0221	0.0081	0.0140	63.23
6	31	374	0.1481	0.0134	0.1348	90.99
7	11	41	0.1794	0.0062	0.1733	96.55
8	34	246	0.1622	0.0213	0.1409	86.89

If a contingency table is constructed, common and different cases can be identified in both configurations. For instance, clusters 2 and 7 in AGNES are identical to those in DIANA.

3.5 Constrained clustering

In certain instances, information regarding the data is available, and this information can be incorporated into the cluster analysis in the form of restrictions. This results in a type of cluster referred to as a constrained cluster. The information can be added as an external variable through the implementation of clustering with an outcome variable. The determination of certain aspects, such as the minimum or maximum size of the clusters, can be attributed to the decisions made by the research.

In some cases, it is necessary to determine whether two data elements must be linked to the same cluster or to different clusters. A methodological

Table 3.4: Contingency table of AGNES vs. DIANA clusters.

	1	2	3	4	5	6	7	8
1	23			7	12			1
2		9						
3	6		52	20	3			
4			43		9	10		
5				127				
6						21		
7							11	
8	3							33

approach has been delineated for investigating the phenomenon of clustering with contiguity constraints in either space or time.

These topics will be explored further in the following Chapter 4.

3.6 Direct partitioning

Direct partitioning is a set of algorithms that divide a dataset into a predetermined number of non-overlapping groups, also known as partitions. As with hierarchical clustering, each point belongs to only one cluster, and the objective is also similar: build clusters with low within-group sum of squares WSS and a high between-group sum of squares BSS. Due to the manner in which clusters are constructed, they are often designated as centroid-based.

K-means, k-medoids (PAM: Partitioning Around Medoids), and CLARA (Clustering Large Applications) are three clustering methods that utilize direct partitioning. K-means and PAM differ in their definition of the center of a cluster. K-means fixes the center of a cluster by computing the average of all points within the cluster. In this manner, the centroid can be defined as an artificial point. PAM employs the medoid (the most central actual data point) as the cluster center, with the centroid representing a real data point. Therefore, k-means is more sensitive to outliers than PAM, since medians are less sensitive. For these reasons, many experts consider PAM to be more robust than k-means. PAM is slightly slower than k-means because it must calculate the distances between all pairs of points in the cluster.

In general, these algorithms are effective when the clusters are spherical. However, identifying clusters with irregular shapes, elongated shapes, or different densities can present a challenge. The most common distance employed in this context is the Euclidean distance.

There are two common elements shared by these approaches that require the researcher's involvement. The first point to consider is that, in contrast to hierarchical methods, the number of clusters must be determined in advance.

The aforementioned tools can also be used to assess whether the selected number of clusters is sufficient to capture the inherent structure of the data.

The second point relates to the positions at which the centers of the selected k clusters are initialized. When the variables used are identical to those in a previous study, it is possible to use the endpoints of the variables in each of the groups as initialization points in the new study.

In text analysis, the words acting as variables may vary from one analysis to another. The same problem occurs with the factors of a CA if they have been used as data input. As a result, the option to establish the centers during the preliminary phase of the clustering analysis is of limited application. In certain instances, the initialization phase may be associated with a random factor, which can yield divergent results, particularly if the data structure is not sufficiently robust. Consequently, it is usually advisable to establish the same seed. In each iteration, the objective is to maximize BSS, which, as previously explained, is equivalent to minimizing WSS.

CLARA is a modification of the PAM method. In this approach, the center is not computed with all the points, but rather with a sample of them. This reduces the time required for calculations, but this sample selection may cause some instability in the results.

As demonstrated in Table 3.5, the results of the k-means method reveal that the number of documents within each cluster is more homogeneous than in the AGNES and DIANA methods. The smallest cluster contains 27 documents with few words, only 81 occurrences. The largest cluster contains a total of 83 documents, which is notably less than the 127 observed in the largest group with AGNES, and the 154 recorded in the cluster with DIANA.

Table 3.5: Intra-cluster results for documents using k-means with 8 clusters.

Cluster	Docs	Ocurrences	ss.tot	ss.intra	ss.between	%Explained
1	54	630	0.0553	0.0159	0.0394	71.32
2	83	557	0.0897	0.0099	0.0797	88.92
3	73	669	0.0384	0.0052	0.0331	86.33
4	35	353	0.2662	0.0495	0.2167	81.41
5	33	369	0.0090	0.0035	0.0055	61.05
6	39	431	0.0517	0.0172	0.0345	66.75
7	46	288	0.3418	0.0881	0.2537	74.22
8	27	81	0.0305	0.0011	0.0294	96.35

The **Xplortext** package provides a solution to the challenge of processing large document sets. This feature enables users to fix a higher number of clusters in k-means than desired, thereby reducing the time required for computing. Additionally, it implements the AGNES method exclusively for the outcome of this k-means clustering. Dendrogram analysis can assist in determining the appropriate number of clusters. For instance, Figure 3.8 shows the dendrogram obtained for the eight clusters generated using the k-means method, and summarized in Table 3.5. Given the close proximity of Clusters

3 and 5, as well as 1 and 7, it would be prudent to explore whether selecting six clusters would be a more effective approach.

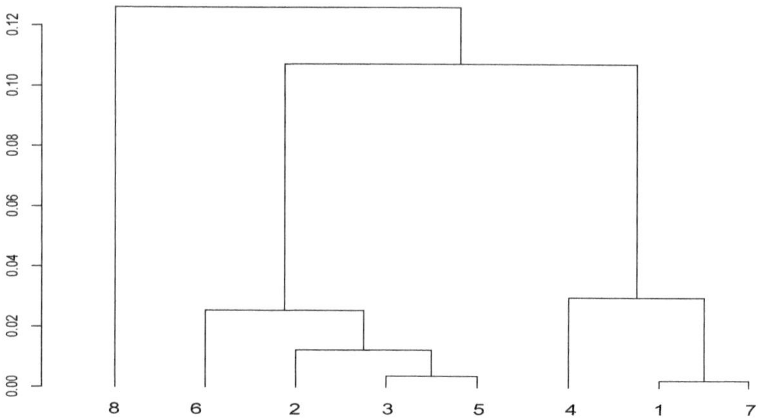

Figure 3.8: AGNES dendrogram from k-means clusters.

3.7 Density estimation

Density estimation are nonparametric clustering methods, although they have a probabilistic background. The objective is to accurately identify the clusters with regions of high density, enabling the precise assignment of one point to the density region that attracts it.

As outlined in Ester's (1996) seminal article on large spatial databases with noise, the concept of *density* refers to the number of points located at a given distance, not to a probability distribution. Each cluster exhibits a density point that is considerably higher than the surrounding area. In areas where noise is present, the density is lower than in any of the clusters. On the other hand, regions of low density can often be spurious and unstable.

It has been asserted that density methods have greater flexibility in identifying the shapes of clusters than parametric models (or model-based approaches). However, this flexibility can sometimes result in the merging of observations that are far apart.

The objective is to obtain a representation formed by connected subsets in a Euclidean space with a density greater than a specified threshold. The fundamental concept is that for each point in a cluster, the neighborhood within a specified radius (eps-neighborhood) must contain a minimum number of points defined in advance (MinPts). There are two types of points in a cluster: *core points*, which are near the center and satisfy the condition that there are MinPts points at a distance of eps, and *border points*, which are far from the center. A *directly density-reachable point* is a point that is in the eps neighborhood of a core point and does not necessarily satisfy MinPts, so it can be a point in the cluster that does not have enough density to be a core point. A point P is *density-reachable* if it is directly density-reachable from the previous point in a chain, where it is not necessary for P to be in the neighborhood of the core point, but it can be part of a connection to it through other intermediate points. Two points P and Q are considered symmetrically density-connected if there is a core point from which both P and Q are density-reachable. This means that P and Q are density-connected, belonging to the same high-density cluster. A *noise point* does not belong to any cluster.

DBSCAN (Density-Based Spatial Clustering of Applications with Noise) is a method whose objective is to identify clusters and eliminate noise without setting the number of clusters in advance (Hahsler et al., 2019). It is true that the number of clusters is not predetermined, but the parameters eps and MinPts must be established, as they determine the number of clusters. So, the problem is not easily solved. For instance, the literature contains recommendations to set the MinPts parameter to twice the number of dimensions. The **dbscan** package provides a method for determining the value of eps. The KNNdistplot function plots the sorted distance of each point to the k nearest neighbor. As the distance to the k-th nearest neighbor increases, the need for a higher eps is suggested. The optimal eps value is typically chosen at the *knee* of the curve, where the distance suddenly increases, indicating a shift from dense areas to sparser noise.

Script **chap3.d** utilizes the 390 answers from the *Health* database that are not empty. The KNNdistplot function is applied using only the first two coordinates from CA, representing k (corresponding to MinPts for DBSCAN) from 4 until 10 (Figure 3.9). The knee for four MinPts indicates an eps value of 0.45. For k equal to 10, eps takes a value of 0.7.

For the parameters MinPts=4 and eps=0.45, eight documents are unclassified, 375 are in cluster one, and the remaining seven are in cluster two. The representation in the factorial plane (Figure 3.10) shows the eight unclassified documents with the symbol "x".

For the parameters MinPts=10 and eps=0.7, eight unclassified documents are obtained, while all the rest are found in cluster one. Given that the KNNdistplot function did not provide adequate partitioning results in this case, a grid was constructed to analyze possible combinations of MinPts and eps. The dbscan function of the **fpc** package was applied to this grid. The objective of this analysis was to reduce noise and identify a greater number of clusters. The eps values analyzed above indicate a very small number of clusters. Only a high number of clusters is detected for low eps and MinPts values, represented in dark tones in Figure 3.11.

To determine if adjusting the parameters would enhance the cluster analysis outcomes, eps=0.08 and MinPts=6 were set. The result showed the

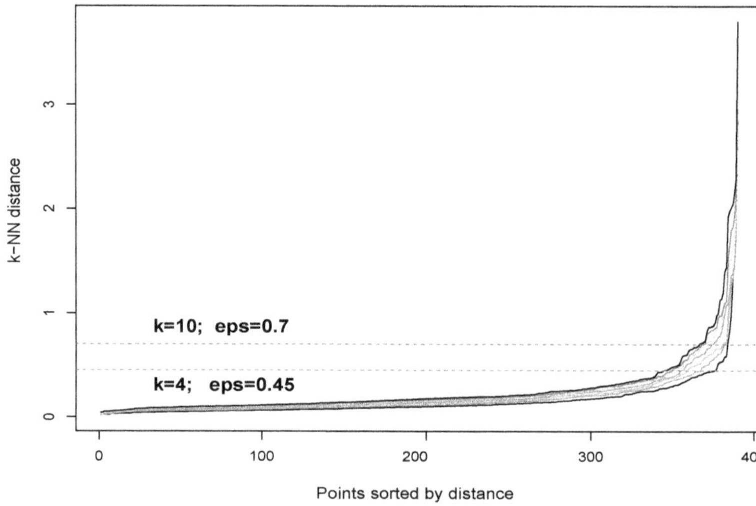

Figure 3.9: DBSCAN. Eps choice for nearest points from 4 to 10.

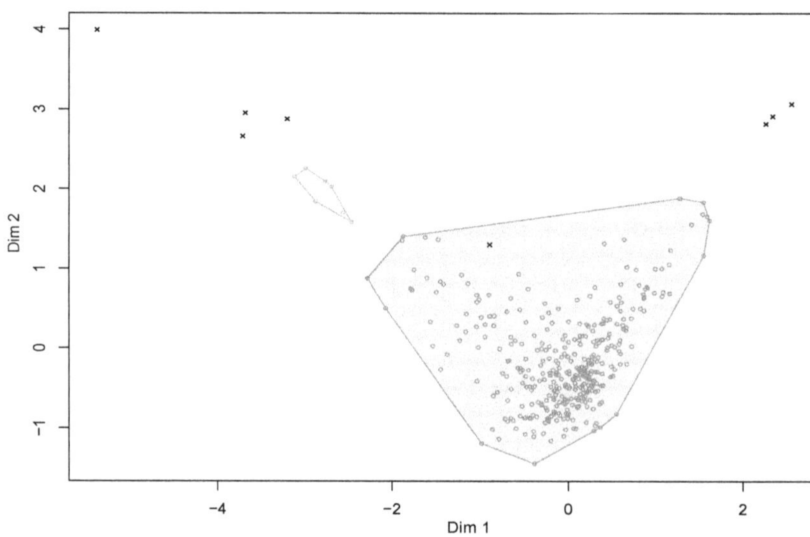

Figure 3.10: DBSCAN clusters for MinPts=4 and eps=0.45.

existence of 224 unclassified documents and seven clusters, one grouping 111 documents, the second 20, and the rest fewer than 10 documents each.

OPTICS (Ordering Points To Identify Clustering Structure) is another cluster density method implemented in the **dbscan** package. When MinPts is set to five, the optics function provides a reachability plot with an interpretation similar to KNNdistplot, suggesting the value eps=0.4. The

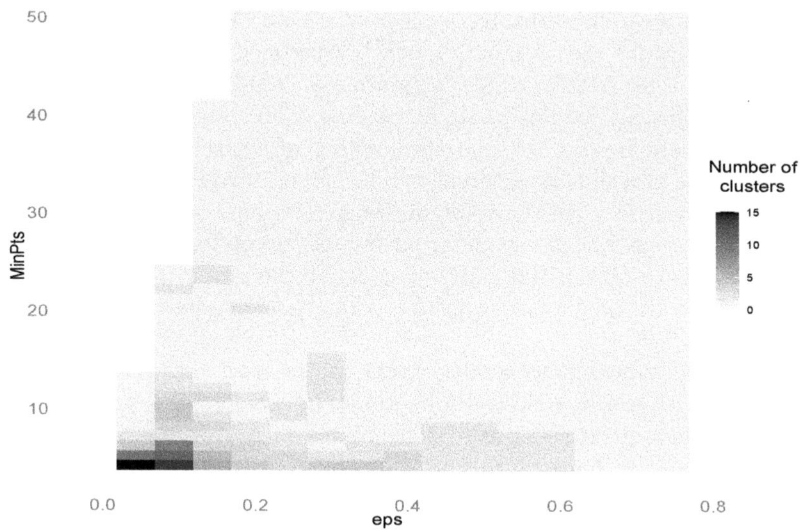

Figure 3.11: DBSCAN grid. Number of clusters by MinPts and eps.

extractDBSCAN function provides one cluster containing 365 documents and two clusters containing six documents each. There are 13 noise points.

The hdbscan function (hierarchical DBSCAN) was applied for MinPts set to five. The analysis provided 23 clusters, with the largest containing 14 documents, and 216 documents were not classified.

Finally, the extractXi function was applied to a wide range of xi parameters. Unfortunately, this resulted in no relevant classification findings.

3.8 Conclusion

Cluster analysis is a collection of unsupervised classification techniques used to discover similarities between documents or words. A variety of forms in which input data can be used have been considered, including DTM, ALT, co-occurrence matrix, the coordinates from a CA, and embedding data.

The number of techniques that fall under the cluster designation is exceptionally high. An attempt has been made to present the main ones in an orderly manner according to the objectives set.

The process of clustering textual data typically involves the classification of a large number of points (documents or words) in a high-dimensional space (words or documents). In this environment, cluster validation shares some aspects of non-textual data analysis, but the process becomes more complex.

The present chapter delineates a range of Cluster Validity Indices, which are used to determine the optimal number of clusters through the application of rules. In addition, relative cluster validation is used to assess the clustering structure by adjusting various parameter values. These include the number of input data factors from a CA and the number of clusters without *a priori* knowledge. The concept of external validation is introduced to assess the similarity between two cluster configurations. The final section of the cluster validation process addresses internal textual characterization validation. The objective is to obtain the labels that define the groups, either from the metadata of the documents or from the words that characterize each of the groups.

Although the same cluster analyses that can be used with documents can be performed with words, BERT (Bidirectional Encoder Representations from Transformers) is applied using word embeddings.

The results of two hierarchical methods are compared for classifying documents: one agglomerative (AGNES) and the other divisive (DIANA).

Within the direct partitioning methods, a k-means analysis is performed, and the results are compared with those obtained from hierarchical methods.

The last type of cluster analysis discussed in this chapter is the density-based methods. Specifically, DBSCAN (Density-Based Spatial Clustering of Applications with Noise), OPTICS, hierarchical DBSCAN, and extractXi were used. Unfortunately, the results of these density methods were not adequate in this case to extract the group structure in the first factorial plane.

4

Constrained clustering defined by the contextual data into the analysis

4.1 Introduction

The traditional view of cluster analysis is located in Unsupervised Machine Learning (UML) methods, which organize objects (data points) into groups based on their similarities or distances with the goal of discovering some unknown underlying structure. In this way, searching for homogeneity within clusters and heterogeneity between clusters is an exploratory data analysis method where labeling the objects can be the way to understand and reveal the structure of the data.

Sometimes the grouping process takes on the character of Semi-Supervised Learning (SSL), which combines the analysis of the data structure with some external information that does not fit into either supervised or unsupervised learning.

This information may also come from researcher's feedback based on previous cluster analyses, which is incorporated into the clustering algorithm in the form of constraints. However, it is usually easier to incorporate these restrictions than to assign the correct labels to each point.

Chapter 1 analyzed a simple example of a constraint when studying the distance between two words, including a restriction based on whether the words belonged to the same grammatical category, as well as the addition of a link constraint, such as "noun + verb."

Establishing a classification of SSL methods is not an easy task, as they depend on many factors, such as objectives, types of variables, and data structure. For a review, see González-Almagro et al. (2025) and Basu et al. (2008). In this chapter, we will consider the following constraints applied to cluster analysis:

- *Clustering with an outcome variable* explores the relationship between clusters while they are forming and a specific known outcome variable or expert knowledge (e.g., grouping of books taking into account the classification of subjects provided by a publisher, the name of the journal of an analyzed text, the age of the respondents in a survey, among others). It is a mixture of exploratory and predictive approaches. The unlabeled internal data is

DOI: 10.1201/9781003397472-4

influenced by the labeled external data. Examples of these techniques include *bayesian profile regression* and *outcome-driven cluster analysis.*

- *Clustering based on size constraint type.* In some cases, it is possible to impose constraints on the cluster structure, such as the minimum or maximum size (number of points) that a cluster should have. Points that do not belong to a valid cluster can be considered outliers. Finding an optimal solution can be computationally expensive, especially when dealing with large datasets. Iterative procedures can be implemented by adapting hierarchical or k-means to introduce this constraint. When a partitioning algorithm is used, the resulting clusters can be further analyzed. For instance, a new hierarchical clustering can be applied to the cluster representatives (their centroids or medoids) to create a dendrogram that visualizes the relationships between the clusters.

- *Link constraints* can specify if two data must belong to the same (*must-link*) or different cluster (*cannot-link*). The goal is to find a result that satisfies as many constraints as possible. Pairwise constrained clustering by *local search* can be very time consuming because it has to calculate the distances between all pairs. However, the k-means approach (see **conclust** package in **R** (Hiep et al., 2016a)) provides faster solutions, although different starting points or seeds may result in different configurations.

- *Soft constraints.* This approach incorporates constraints with flexibility. Instead of being rigid rules, constraints are treated as penalties. In the event that a constraint, such as a must-link, is violated, a cost or penalty is added to the solution. The algorithm then tries to identify a clustering that minimizes the total penalty, allowing it to violate some constraints if it results in a significantly improved overall structure.

- *Clustering with contiguity constraints in space or time.* This method is employed when clusters must be spatially or temporally connected.

  The algorithm typically takes two inputs:

  1. A dissimilarity matrix **D** for the features (e.g., demographic data).
  2. A contiguity matrix **C** for the structure (e.g., defining which objects are neighbors).

  This contiguity matrix **C** acts as a *hard constraint.* For example:

  - In a spatial constraint (e.g., a map), $C(i, j) = 1$ if areas i and j share a border, and 0 otherwise.
  - In a temporal constraint, a point t_i is only considered contiguous with its immediate neighbors, t_{i-1} (the previous point) and t_{i+1} (the next point).

The algorithm is then restricted so that a cluster is only valid if all its members form a single connected component.

The corpus frequently comprises a series of documents that are related in some way, either according to their date of production, as diachronic studies, the gender or age of the respondent in the case of open questions, or some other variable (quantitative or qualitative).

Hereafter, *contiguity* may be understood as a generic chronological or spatial relationship between the documents or as an ordered *aggregated lexical table* (ALT) built from the ordered categories of a qualitative variable.

In this chapter, we will focus exclusively on the grouping of documents, not the clustering of words. With a pragmatic objective in mind, the present study starts with the description of the corpus. This corpus will serve as an example for the implementation of different types of constrained clustering.

4.2 Application on a corpus of Spanish investiture speeches

4.2.1 The corpus

As an example of applying a cluster analysis with time constraint and stop rules, the corpus *SpanishDisc18*, which contains the 18 inaugural speeches of the Spanish prime ministers (1979-2023), has been used. After the Franco dictatorship, which lasted from the end of the Civil War (1936-1939) until his death (November 20, 1975), democracy returned to Spain. A new Constitution was approved on December 6, 1978, which stipulated that the candidate for prime minister, proposed by the King of Spain, must be elected by Parliament (with an absolute majority in the first ballot, but only a simple majority in the second). In each investiture speech, the candidate presents his program of government. Then, the deputies vote on whether or not to grant their confidence.

From 1978 until 2025, a total of 18 speeches were delivered by eight different politicians: Adolfo Suárez and Leopoldo Calvo-Sotelo, both of Unión del Centro Democrático (UCD, center right), Felipe González from Partido Socialista Obrero Español (PSOE, left), José-María Aznar from Partido Popular (PP, right), José-Luis Rodríguez Zapatero (PSOE), Mariano Rajoy (PP), Alberto Núñez Feijóo (PP) and Pedro Sánchez (PSOE). They are represented in the graphs by their rank (from 1 to 18), two letters summarizing the name of the candidate and the last two indicating the year.

The Spanish governments have demonstrated a notable degree of stability, as only two legislatures have featured more than one investiture speech: the first, due to Adolfo Suárez's resignation at the beginning of 1981, and the last, where Núñez Feijóo was not elected. It should be noted that the second ballot for electing Leopoldo Calvo-Sotelo, who did not have an absolute majority on February 19th, was abruptly interrupted by the military coup. Finally, after the unsuccessful coup d'état, Leopoldo Calvo-Sotelo was elected on February 25th with an absolute majority.

The 18 inaugural speeches have been taken from the version published in the *Diario de Sesiones del Congreso de los Diputados* (Journal of Sessions of the Congress of Deputies).

4.2.2 Preprocessing

Regarding the preprocessing of texts, several normalization steps were performed. First, capital letters at the beginning of sentences were converted to lowercase, but those used to differentiate homographic words were retained. Therefore, the term *Gobierno* (the government) is distinguished from *gobierno* (I govern). It will therefore be necessary to specify in the script that we wish to retain case sensitivity. Their writing has been unified (e.g., *IAE* and *Impuesto de Actividades Económicas*, Business Activities Tax) and multiword entities were joined with an underscore to represent them as single units (e.g., *Seguridad_Social*, social security). However, other common procedures like stemming and lemmatization were not employed in this study.

The corpus used is in Spanish, which is the language used for the analysis. The results have been translated into English.

The TextData function of the **Xplortext** package (Bécue-Bertaut et al., 2024) is used to retrieve the original text of the source documents and to apply some preprocessing rules before building the DTM matrix (script in **chap4.a.R**). The custom stopwords defined by the user are as follows: *consiguiente, ello, hacia, punto, Señorías, si, Sus, vista, A, B, C, D, E, F, a, b, c, d*. This list includes expressions often repeated in a single speech as a verbal tic, such as, for example, *por consiguiente* (therefore), or as enumeration markers (*A, B, I*, etc.). In addition to this custom list, the standard Spanish stopwords from the **tm** package (Feinerer and Hornik, 2025) and all numbers are also removed.

Only words with a frequency greater than nine and appearing in at least two speeches are retained.

Before selection, the corpus size is 188,497 total occurrences and 13,628 unique words. The average speech length is 10,472 occurrences.

After selection, the corpus contains 67,056 total occurrences and 1,941 unique words. The average speech length is now 3,725 occurrences.

The focus here is on presenting the approach taken and the choices made, rather than an in-depth interpretation of the results. However, words are used to characterize some of the clusters obtained.

4.3 Correspondence analysis and supplementary elements

As described in Chapter 2, CA is a powerful tool for visualizing the main relationships between documents and words by analyzing the first factorial axes and removing the last factorial axes belonging to the random structure. CA results can be enriched by incorporating a constrained variable as a supplementary variable.

If the documents are closely related to the vocabulary, the first axes reflect the evolution of the documents. However, measuring the importance of the constraint in the corpus structure can be challenging when the number of axes is high (large number of documents and words), since the restriction can be distributed over several different factorial axes. In such cases, exploring the use of different axes to analyze the grouping of the data with the contiguity constraint may be beneficial.

In the context of political speeches, the corpus can be related to a contextual variable, such as chronology or political party.

When the vocabulary is highly related to order, CA can be used to visualize the document trajectory on the first factorial axes. This helps to explain the variability in the vocabulary.

After conducting the CA, the following five axes have been retained. These correspond to the above-average eigenvalues (0.0634), which are 0.176, 0.094, 0.089, 0.082, and 0.070, respectively. All together, they keep 47.4% of the total variance (16.34%, 8.71%, 8.27%, 7.61%, and 6.45%, respectively).

It has been decided to use chronology as the supplementary variable, rather than individual years, as there are several investiture speeches that have taken place in the same year, such as 12.SA16, 13.Rj16, and 14.Rj16 in 2016, and 17.Fj and 18.Sa23 in 2023.

Table 4.1 shows the statistics for the variable *chronology* on the first five factors. It is noteworthy that there is a strong correlation with the first factor (coordinate = −0.945). The cosine squared (the squared correlation) demonstrates that this first factor alone accounts for 89.28% of the chronology variable's total variance.

Table 4.1: Statistics considering chronology variable as supplementary in CA.

Chronology variable	Dim 1	Dim 2	Dim 3	Dim 4	Dim 5
Coordinates	−0.945	−0.007	−0.047	−0.140	−0.199
Cosinus squared * 100	89.28	0.01	0.22	1.96	3.97
Acumul. Cosinus squared * 100	89.28	89.29	89.51	91.47	95.44

As illustrated in Figure 4.1 (first row, right), the strong chronological relationship in the first factor is evident, with a proximity to the maximum value of the circle of correlations. The figure on the left further illustrates this, revealing the temporal progression of the discourses from the right part of the first factor to the left part. It is reasonable to conclude that, given the evident relationship between the chronology and the first factor, there is negligible correlation between the chronology and the orthogonal subsequent factors.

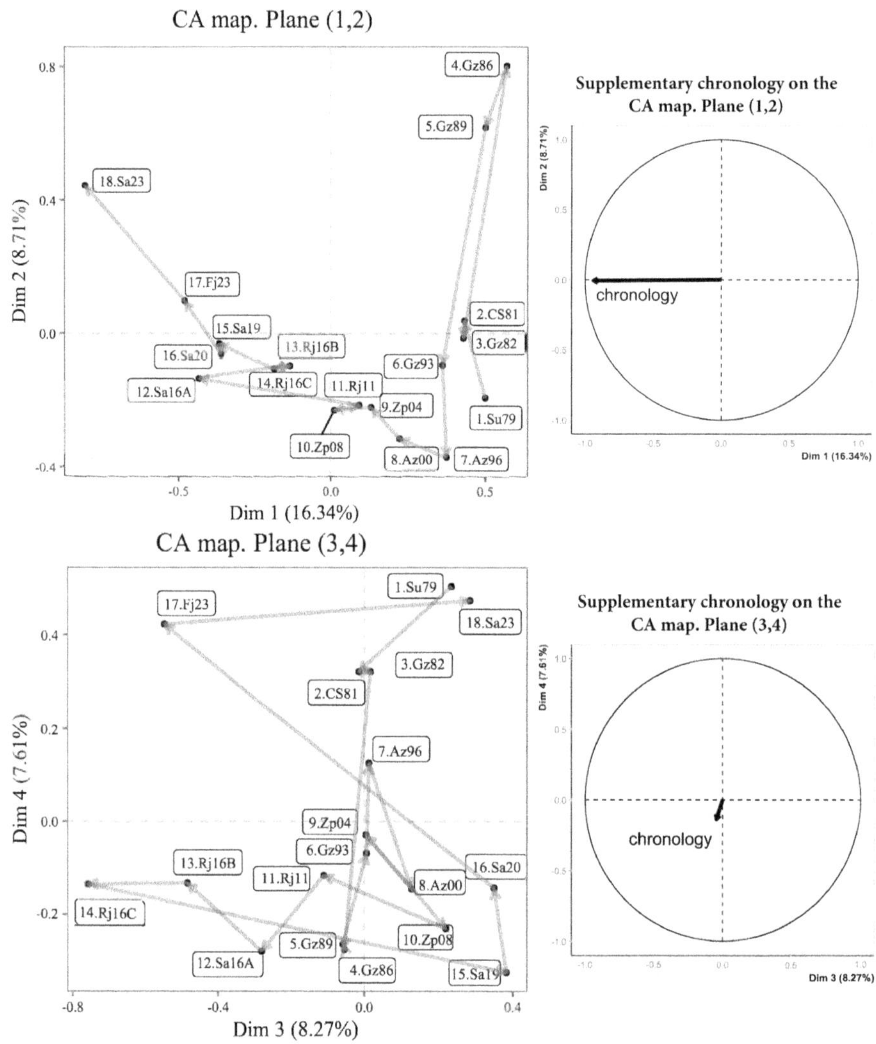

Figure 4.1: Representation on the CA planes of documents (left) and chronology as supplementary variable (right).

An examination of the first factorial plane depicted in Figure 4.1 reveals the initial period (1981-1989), predominantly characterized by terminology associated with the transition from dictatorship to democracy, the Socialist

Party's ascent to power, the structural organization of the State of the Autonomous Regions, foreign policy, Spain's consolidation within the European Union, and its membership in NATO.

The second axis shows the developmental trajectories of these first seven speeches, while the first axis depicts minor alterations up to the 7.Az06 point. Regarding the political party, which is used as a supplementary qualitative element, the first factor is characterized on the positive side by UCD with a test value (vtest) 42.8 and on the negative side by PSOE with vtest −25.5. The second factor is characterized on the positive side by PSOE with vtest 32.4 and on the negative side by PP with vtest −28.7 and UCD with −9.0.

From 7.Az06, there is a shift towards the negative part of the first factor and the positive part of the second. However, the chronology only shows a relationship with the first factor.

From 4.Gz86 to the last speech 18.Sa23, a clear *horseshoe* pattern (also called *bow*, *arc* and *Guttman* effect) is identified (Benzécri (1973); Chavent et al. (2018); Lebart et al. (1998, Chapter 7)), which suggests that the analysis is dominated by the presence of a gradation or ordination relationship in the structure of the documents, since the words appear, are maintained over time and sometimes also disappear according to the problems and solutions that are exposed in this type of corpus. At this stage, it is important to remember that the order of the documents acts, in this study, as a supplementary element, and that it is independent of the results obtained in the CA.

The representation of the discourses in plane 3-4 does not present a relationship with the chronology. Factor 3 contrasts the speeches of the PP in the negative part with those of the PSOE and the UCD in the positive part. In factor 4, UCD is in the positive part, far from the PP and PSOE in the negative part. There are some differences between the PP and PSOE within the negative part.

The first factorial plane depicted in Figure 4.1 shows the documents arranged in a parabola (from 4.Gz86 to 18.Sa23). The points at the beginning of the timeline (the first seven speeches on the right) are more concentrated, while those from the middle of the period tend to be more separated.

This pattern is typical when the vocabulary of the speeches evolves over time, or when an ordinal relationship explains the structure of the documents, as is common in ecological studies, where site values change along an environmental gradient (Legendre and Legendre, 2012).

If the first axis is folded, an arc can be constructed by joining the extremes (18.Sa23 with 4.Gz86 and 5.Gz89). Consequently, the second axis can be calculated as a quadratic function of the first axis, the third as a cubic function of the first, and so on (Hill, 1974).

It is common to find some speeches with high coordinates on the second axis and, logically, more moderate positions in general terms over the rest. This leads to a problem of interpretability. The second axis is conditioned by the results of the first axis, since they must be orthogonal.

Figure 4.2 illustrates that the arc effect is also evident in the first factorial plane representing words, with a limited number of characteristic words at the extremes and a substantial number grouped together in the middle.

It is important to bear in mind that the coordinates of speeches can be obtained from the weighted coordinates of words, and vice versa. In the factorial plane (3,4) the Guttman effect is not observed. To validate the trajectory, a bootstrap was performed on the occurrences (Valencia-García and Alvarez-Esteban (2012); Alvarez-Esteban et al. (2010); Chapter 2 of this book). For each speech, small regions were obtained that did not tend to overlap, indicating that the positions of each speech and of the trajectory itself have a high degree of stability, which facilitates the interpretation of the results.

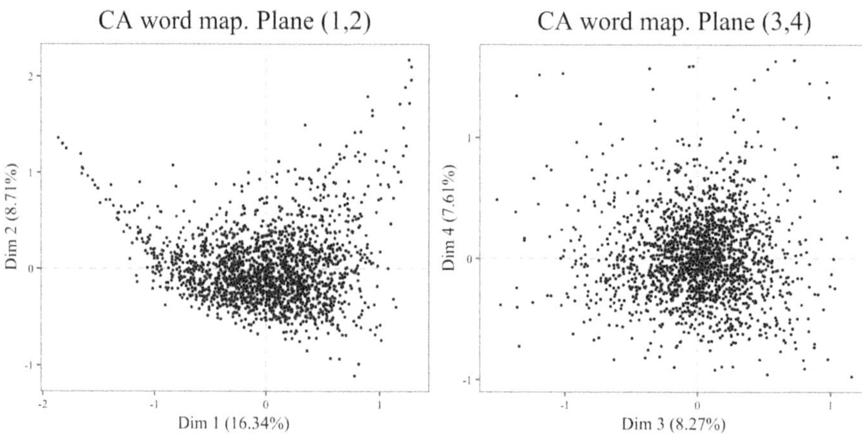

Figure 4.2: Representation on the CA planes of words.

Clustering on the CA scores of the speeches allows the documents to be grouped together based on these underlying relationships. To facilitate data clustering, the analysis of the departures from this arc pattern can be meaningful and constitute relevant information. To this end, one possible approach might be to eliminate or analyze this arc effect before grouping the discourses. In this chapter, we present several techniques based on CA, using the 18 discourses described above as an application that allows us to compare the results obtained.

4.4 Hierarchical cluster analysis from CA coordinates

Combining factorial techniques with hierarchical cluster analysis (HCA) enriches the visual representation of the former by positioning clusters on factorial planes and examining their relationship with the original and latent variables. The coordinates of the first axes obtained from CA can be used as quantitative input data when performing cluster analysis, bearing in mind that the factors are centered and that the dispersion of individuals in each factor is related to the eigenvalue. Therefore, the axis should not be typified if we wish to maintain the importance of each factor when performing cluster analysis.

There are a variety of cluster analysis methods that can be used with different statistical distances. In this chapter, we will first perform an unrestricted *agglomerative hierarchical clustering* (AGNES, acronym for Agglomerative Nesting) with the Euclidean distance and the Ward.D2 method obtained from the factor coordinates of the first five axes. In the following sections, we will include restrictions and compare the results (script in **chap4.b.R**). It is important to note that restrictions in HCA can lead to inversions when constructing the dendrogram.

Figure 4.3 (left) shows this dendrogram. The observed relationships of proximity between the speeches are due to the evolution of the topics and the adaptation to the problems related to the reality of the moment.

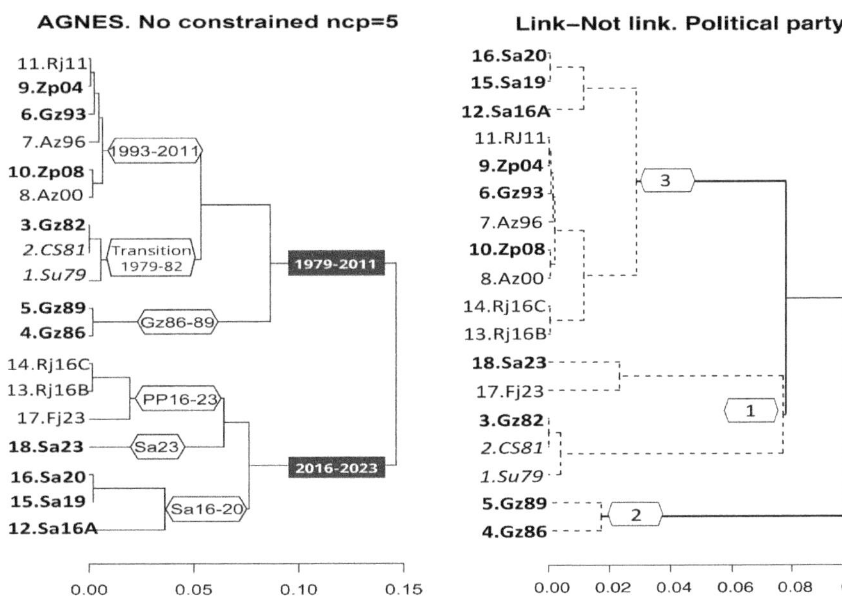

Figure 4.3: Dendrograms without constraints (left) and Link – Not link constraints (right). UCD in italic, PSOE in bold and PP in normal font.

The selection of six clusters led to the identification of two predominant periods, with a delineation of the cutoff point occurring in the year 2016.

The first group (1979-2011) is divided into three subperiods:

- 1979-1982. The transition in Spain from Franco's dictatorship to a democratic regime marks the beginning of this shift. This group includes 1.Su79 and 2.CS81, speeches by two politicians from UCD, a center-right party created by Adolfo Suárez that quickly declined after he left political life. It also includes 3.Gz82 by Felipe González from PSOE.

An important theme that emerges is the *freedom* (the most characteristic word in this period), along with *liberty* and *peace*, centering on the *person*: *people*, *citizen*, *man*, *men*, *human*, *Spaniards*, and *solidarity*. A new structure of the *State* and *organization* under the new *Constitution*, development of *Autonomous Regions*, *statutes of autonomy*, *public administration*, and the *legal* (*law*, *laws*, *legislative*) reforms needed to move from a dictatorship to a democratic *regime*.

This change is also intended to facilitate greater openness to the *outside world*, to *Europe*, *European accession*, and *international* integration. The manner in which the change needs to be implemented is important (the fourth most significant word in this period is *politics*, appearing alongside words such as *order*, *action*, *program*, *solution*, and *solutions*).

In the *economy* field, there are words such as *prices* and *inflation* (near 14%, close to EU countries and significantly above the USA), *crisis*, *industrial reconversion*, and *unemployment* (during 1978-1984, more than 20% of employment in *industry* and *agriculture* was lost).

González's inaugural speech was delivered on December 1, 1982, in close temporal proximity to the February 23, 1981 putsch. This speech attempted to maintain continuity in the face of concerns regarding the possibility of a subsequent coup.

- 1986-1989. The second subperiod comprises two speeches delivered by the same author, Felipe González (4.Gz86 and 5.Gz.89). As a result, some individual characteristics of his style of expression may appear (*etc.*, *I would like to*, *change*, and *changes*). These *changes* refer mainly to *economic* issues (*economy*, *modernization*, *infrastructures*, *balance*, *consumption*, *tax rates*, *deficit*, *inflation*, and *recovery*). They also address international *relations* (*European Union*, *Accession Treaty*, *integration*, *Single European Act*, *Community*, *European*, *countries*, and *Mediterranean*). There is a *terrorism* problem (*terrorist*, *ETA*) and internal issues such as the *Penal Code* and *regional autonomy*.

- 1993-2011. The third subperiod includes a wide range of authors, starting with González's fourth speech (6.Gz93), followed by two speeches by Aznar (7.Az96 and 8.Az00), two speeches by Zapatero (9.Zp04 and 10.Zp08), and the first speech by Rajoy (11.Rj11). This covers a period of 18 years.

  The most suitable words to characterize this period are *reform* and *reforms*, along with *improvement* and *improve*, *renewal*, *modernization*, and *innovation*. Discussion of the *State's* structural framework continues, covering both the central level (*constitutional*, *Public Administrations*) and the decentralized levels (*Autonomous Communities*, *Local entities*). At the legislative level, words include: *justice*, *judicial*, *court*, and *law*. The ongoing process of *internationalization* (*international*, *foreign*, and *European*) persists, as do problems related to *employment* (*labor*) and *social* issues (*solidarity*, *immigration*, and *culture*). *Economic* terms are frequent (*economy*, *economic stability*, *budget*, *spending*, *tax*, *information*, *competitiveness*, *competition*, *markets*, *business*, *entrepreneurship*, *telecommunications*, and *technology*).

The second period consists of five speeches over a short period (2016-2023). It is marked by the absence of clear majorities and the challenges in achieving consensus with other political parties. It also contains three subperiods with overlapping speeches.

- The fourth subgroup corresponds to three speeches by the PP (two by Rajoy 13.Rj16B and 14.Rj16C and one by Feijóo 17.Fj23) designed as PP16-23. These were delivered after Sánchez's speech in March 2016 (see 12.Sa16B in the next fifth subgroup of speeches), in which he failed to obtain the necessary votes to be elected president. In August, Rajoy made a first unsuccessful attempt, yielding the same result in the second round of voting in September. However, he was subsequently elected president in October, due to the 68 abstentions from the PSOE in the second round. It is also noteworthy that Feijóo did not secure the presidency in 2023.

  As the candidates did not attain an absolute majority in Parliament, the most frequently used words refer to the votes and the sovereignty of the Spanish people in the elections that yielded a relative majority for the candidate's party.

  A large number of words refer to the search for votes, addressing specific *political parties*, the *Chamber* in general (*support*, *stable*, *parties*, *I thank you*, *here*, *today*, *pact*, *investiture*, *Honorable Members*, *govern*, *government*, *Parliament*, *alternative*, *unity*, *session*, *agreements*, *speaking*, *commitments*, *uncertainty*, and *partners*).

  Words typical of the candidate's vocabulary are also observed: *(I'm going*, *attend*, *I'm coming*, *he says*, *fortunately*, *give*, *can*, *reason*, *reasons*, *I must*, *I propose*, *allow me*), and very few words about the economic and social reality of the country (*millions*, *water*, *taxes*, and *equality*).

- The fifth subgroup consists of three speeches by Sánchez (12.Sa16A, 15.Sa19, and 16.Sa20). As previously mentioned, Rajoy's two speeches from 2016 are classified in the fourth group.

  Sánchez's speeches contain around 200 characteristic words. There are several words related to the way of addressing *Parliament* (*Honorable Members*, *Parliamentarian*, *Government*, *Rajoy*, and *female members*), others related to the way of governing (*let's go*, *Socialist*, *progressive*, *move forward*, *agreement*, *change*, *coalition*, *dialogue*, *parliamentary group*, *approve*, *we will approve*, *block*, *we will promote*, *promote*, *transformation*, *comprehensive*, *we will put*, *regeneration*).

A multitude of words have been identified, including those related to human behavior (*people*, *inequality*, *violence*, *gender*, *women*, *discrimination*, *exclusion*, *poverty*, *children*, *childhood*, *gaps*, *citizenship*, *cohesion*, *education*, *schooling*), *labor* (*contracts*, *wages*), the *digital revolution* (*digitization*), *ecological* and *sustainability*, as well as *corruption*.

- Subcluster six consists exclusively of the final speech (18.Sa23), which is characterized by an agreement with *Catalan separatists* to achieve an *amnesty law* that would cover the actions undertaken during the 2017 Catalonia referendum. It relates to aspects such as the *Constitution*, *law*, and *legality*.

  Sánchez aims to present a *democratic left-wing government* (*democracy*, *democracies*), contrasting to Feijóo's latest speech (*Partido Popular*, *right-wing*), who received support from *VOX* (*far-right*).

  Sánchez highlights policy differences (*hatred*, *LGTBI*, *energy*, *ecology*, *climate*, and *climate emergency*), includes words in the feminine gender (*citizens*, *women*, *Spaniards*, *feminism*, *workers*), and uses words that differentiate him from the *parties* that did not support him (*pay*, *tax*, *taxes*, *funds*, *cutbacks*, and *welfare*). He also uses positive words, such as: *let's go*, *moving forward*, *progress*, *continue*, *transformations*, *choose*, *follow*, *reinforce*, and *expand*.

4.5 Clustering based on link constraints

Based on the example of the speeches and the first five coordinates of the CA, it is possible to establish link constraints. The package **conclust** (Hiep et al., 2016a) allows the user to build a list of must-link constraints and another list of cannot-link constraints. The number of clusters can be set by the user or left for the package to calculate automatically. The ccls function uses an algorithm for pairwise constrained clustering by local search, minimizing the clustering cost function using Tabu search (Hiep et al., 2016b).

In this example, the constraint is constructed based on the political party to which the speaker belongs (PP, PSOE, or UCD). The first step is to create the must-link constraints for each group, defining a data frame with two columns (script **chap4.c.R**). Each row contains two points that must appear in the same cluster. Once the data frames for all the groups have been built, they are combined. Similarly, it is possible to define a new data frame with the cannot-link restrictions.

Using the automatic selection of the number of clusters in the ccls function, 10 clusters are proposed. This indicates that membership of a political party is not closely related to the speeches delivered by the other candidates of their political party, as can be seen in Table 4.2.

Table 4.2: Without constraints (WCns), link constraints with automatic selection (10 clusters) and 3 clusters. Spch: UCD in italic, PSOE in bold and PP in normal font.

Spch	*1*	*2*	**3**	**4**	**5**	**6**	7	8	**9**	**10**	11	**12**	13	14	**15**	**16**	17	**18**
WCns	1	1	1	2	2	1	1	1	1	1	1	3	3	3	3	3	3	3
Cl10	1	2	2	3	3	4	4	5	4	5	4	6	7	7	8	8	9	10
Cl3	1	1	1	2	2	3	3	3	3	3	3	3	3	3	3	3	1	1

Once the 10 clusters were built, the number of each cluster (which is arbitrarily labeled) was modified to facilitate interpretation. It was observed that only chronologically consecutive groupings of the same authors were produced for speeches: 4.Gz86 and 5. Gz89 by González (PSOE) (which were very different from the rest in previous analyses), speeches 13.Rj16B and 14.Rj16C by Rajoy (PP), and 15.Sa19 and 16.Sa20 by Sánchez (PSOE).

Setting the number of clusters to three (equal to the number of political parties) makes it easy to see how the clusters demonstrate a clear chronological component that has a greater influence than the candidate's political party.

The characteristic cluster formed by the two central speeches of González (4.Gz86 and 5.Gz89, coded as 2 in Table 4.2) is obtained again.

Regarding the other two groups, the first three speeches, corresponding to the transition (two from the UCD and the first from González, coded as 1) are grouped with the last two (one from Feijóo PP and one from Sánchez).

The third cluster (speeches coded as 3) contains discourses from six to 16, alternating six times between PSOE and PP speeches, which again indicates the greater importance of the chronological component over the candidate's political party.

The adjustedRandIndex (ARI) function (Rand, 1971; Hubert and Arabie, 1985) of the **mclust** R package (Scrucca et al., 2023) is a robust method to compare two vectors containing the outcomes. It measures the amount of agreement between the two cluster partitions, correcting for the agreement expected by chance. ARI can be negative if the index is lower than that expected by random comparison, although the index typically falls within the range $[0, 1]$. An ARI value of 0 indicates that the two cluster partitions do not agree beyond chance, while an ARI value of 1 indicates that both results represent the same clusters.

A comparison between the results of the constrained and unconstrained clustering methods yielded an ARI value of 0.2, indicating a low degree of agreement between the two resulting partitions. This suggests that the introduction of the restriction has indeed modified the classification structure. Similar results (low agreement) were obtained using other indices that allow for comparison of dendrograms, such as the silhouette (0.20), cophenetic correlation (0.33), and Baker's gamma correlation (0.28).

To graphically compare the results with the unrestricted cluster, Figure 4.3 (on the right) contains an artificially constructed dendrogram with the restrictions of the political parties. After determining the cluster membership for each point, three clusters are formed using the Ward.D2 method. These three clusters allow for the determination of the corresponding centroids. With appropriate weighting, a new dendrogram can then be generated using Ward.D2, containing only three labels, one for each cluster. Subsequently, the sub-dendrograms are incorporated by adjusting their height at the corresponding nodes on the global dendrogram. It should be noted that inversions may occur using this procedure. As can be seen, only the group marked with the number 2 contains speeches from the same political party (PSOE), once again highlighting the importance of chronology over political party.

The left side of Table 4.3 presents a confusion matrix, showing the diversity in the distribution of clusters when their number is automatically set to 10. This automatic selection results in a maximum of two speeches in each cluster.

By constraining the analysis to three clusters to match the number of political parties (right side of Table 4.3), a confusion matrix is obtained in which the two UCD speeches are coded together as 1. However, this cluster also includes the PP speech 17.Fj23 (Feijóo), the first from González (3.Glez82), and the last from Sánchez (18.Sa23). Cluster 2 contains the two central speeches of González (4.Gz86 and 5.Gz89), which are also grouped separately from the rest. Cluster 3 includes all the speeches between 6.Glez93 and 16.Sa20, corresponding to five speeches by the PP and six by the PSOE.

Table 4.3: Summary of number of speeches using Link constraints with automatic selection (10 clusters) and 3 clusters.

	10 clusters										3 clusters		
Cluster	1	2	3	4	5	6	7	8	9	10	1	2	3
PP				2	1		2		1		1		5
PSOE		1	2	2	1	1		2		1	2	2	6
UCD	1	1									2		

In summary, we can highlight that the construction of clusters based on must-link/cannot-link constraint criteria can provide results that differ greatly from the initial criteria if there is a structure that can better explain the differences between the vocabulary of different individuals. Additionally, it is also possible to analyze the variation in the composition of each cluster obtained under the constraints or without those constraints.

4.6 Clustering based on size constraint

Next, as an example, we will ensure that each of the obtained clusters contains a minimum of three discourses. This is not the case for the original unconstrained clusters (WCns in the first row of Table 4.4), where cluster

2 contains only two discourses (4.Gz86 and 5.Gz89). This constraint can be useful for avoiding outliers in the final solution.

We will consider including the size constraint, starting with AGNES, and two variations of the divisive analysis method (DIANA).

4.6.1 Minimized pairwise distances with size constraint

Implementing this process utilizes the **ssclust** package (Higgins et al., 2016; Savje et al., 2024), which uses a function called sc_clustering that minimizes within-cluster dissimilarities while complying with the chosen minimum size constraint. Additionally, it also allows for the imposition of additional constraints on the composition of each cluster. The function applies an algorithmic process that initially generates a representation of the distances between data points in a sparse graph (script **chap4.d.R**). In the context of graph theory, each data point is represented by a vertex in the graph, and the arcs are weighted by the distance between the points they connect. Consequently, a set of vertices in the graph is selected to act as "seeds," with the objective of minimizing the arc weight. The initial clusters are constructed from the seeds' respective neighborhoods, with the stipulation that they do not overlap. According to this premise, each local neighborhood in the graph must contain a number of vertices equal to the size constraint. This ensures that the size constraints are respected. Finally, the remaining data points are allocated to clusters.

The sc_clustering function provides the number of clusters with labels starting from zero. Due to the arbitrary nature of these numbers, they have been recorded in this example in accordance with their chronological sequence, starting from 1.

Table 4.4 shows the results for each of the 18 speeches. The function has grouped them into four clusters (row Size3), ensuring that each group has a minimum of three speeches, showing a clear chronological component.

Table 4.4: Without constraints (WCns4) and 4 clusters, (WCns5) 5 clusters, (WCns6) 6 clusters, size link constrain equal 3 (Size3), divisive clustering from a single cluster (DivS) and divisive clustering from Size3 (Div3). Spch: UCD in italic, PSOE in bold and PP in normal font.

Spch	*1*	*2*	**3**	**4**	**5**	**6**	7	8	**9**	**10**	11	**12**	13	14	**15**	**16**	17	**18**
WCns4	1	1	1	2	2	1	1	1	1	1	1	3	4	4	3	3	4	4
WCns5	1	1	1	2	2	1	1	1	1	1	1	3	4	4	3	3	4	5
WCns6	1	1	1	2	2	3	3	3	3	3	3	4	5	5	4	4	5	6
Size3	1	1	1	1	1	2	2	2	2	2	2	3	4	4	3	3	4	4
DivS	1	2	2	3	3	3	1	1	4	2	5	4	5	5	4	6	6	6
Div3	1	1	1	1	1	2	2	3	3	3	2	4	5	5	4	4	5	5

The increase in the minimum cluster size from two (in the unconstrained configuration, WCns4) to three (Size3 with four clusters) shows a moderate similarity to the unrestricted configuration, quantified by an ARI value of 0.6. This value is considerably higher than the ARI obtained using link constraints for political parties (0.2). The cophenetic (0.97) and Baker's gamma (0.96) values confirm that the two structures are very similar.

According to the results outlined above, it is feasible to construct four dendrograms independently and then combine them into a single dendrogram. This approach enables a comparison of the results with those of previous dendrograms. By looking at the Figure 4.4 and Table 4.4, the changes in structure can be clearly understood from the change in size introduced by the constraint.

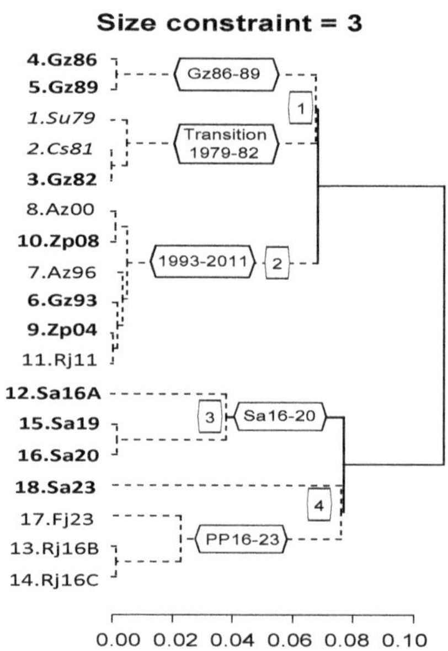

Figure 4.4: Dendrograms with size constraint equal to three, pairwise.

- Cluster 1 contains the first three speeches, corresponding to the Transition, as well as speeches 4 and 5, which were classified separately without restrictions but are now included when the cluster size is increased to three.
- Cluster 2 represents the period from 1993 to 2011. It includes the last speech by González (6.Gz93), when he was not considered the winner in the polls, and his speech was different from his previous ones. Cluster 2 ends with 11.Rj11. Without constraints, this cluster was joined to the three first speeches (1.Su79-3.Gz82). Together, clusters 1 and 2 span the same overall period, both with and without the constraint: 1979-2011.
- Clusters 3 and 4 are identical for the period 2016, both in content and internal structure.

Once again, we see the importance of chronology in this example when constructing clusters. It has also been verified that the size restriction has been respected.

4.6.2 DIANA with size constraint

When the data structure is complex, HCA is usually the preferred option as it is effective at identifying small clusters. DIANA is a top-down clustering method that starts with one cluster containing all the points. The objective is to split the clusters into smaller groups such that the new clusters are as dissimilar as possible. The process finishes when there are as many clusters as points, or when a defined number of clusters is reached. In the latter case, the clusters obtained may differ from those obtained using an agglomerative method (AGNES).

Depending on the grouping method used with AGNES, it is possible to find clusters with a high number of points and high density. DIANA may be recommended when there are a large number of points and an agglomerative method cannot be used, or smaller clusters are desired.

The package **ssclust** has a function called hierarchical_clustering that automatically suggests the number of clusters while respecting the minimum size of each cluster.

The results are shown in the fifth row of Table 4.4, with the label DivS, starting the process from a single cluster. They can be compared with those obtained for the unrestricted AGNES cluster (WCns6) in the third row of this table. Six groups are proposed, which disrupts the chronological structure obtained in previous configurations. This confirms that divisive and agglomerative methods can yield greatly differing results.

The ARI index is 0.11 when comparing this result with the AGNES method for the six clusters. The low cophenetic correlation coefficient (0.38) and Baker's gamma statistic (0.41) suggest significant differences between the two configurations for six clusters.

Figure 4.5 (left) shows the dendrogram for the six clusters obtained from DIANA, which was constrained to include a minimum of three speeches in each cluster. The three clusters at the top of the dendrogram (labeled 1, 2, and 5) contain two UCD speeches, five PP speeches, and two PSOE speeches.

Only a few groups of speeches are clustered based on when they were delivered. These groups belong to the same speaker: Aznar (7.Az96 and 8.Az00) and Rajoy (11.Rj11, 13.Rj16B, and 14.Rj16C). It is noteworthy that these chronologically grouped speeches are not close to other speeches in time.

The group of speeches by González (4.Gz86, 5.Gz89, and 6.Gz93) is clearly differentiated from the rest in the central part of the dendrogram (label 3), and is far away from his earlier speech (3.Gz82, label 2).

Among the six speeches at the bottom, we find all four speeches by Sánchez and one by Zapatero (five PSOE speeches total), compared to only one speech by the PP (17.Fj23).

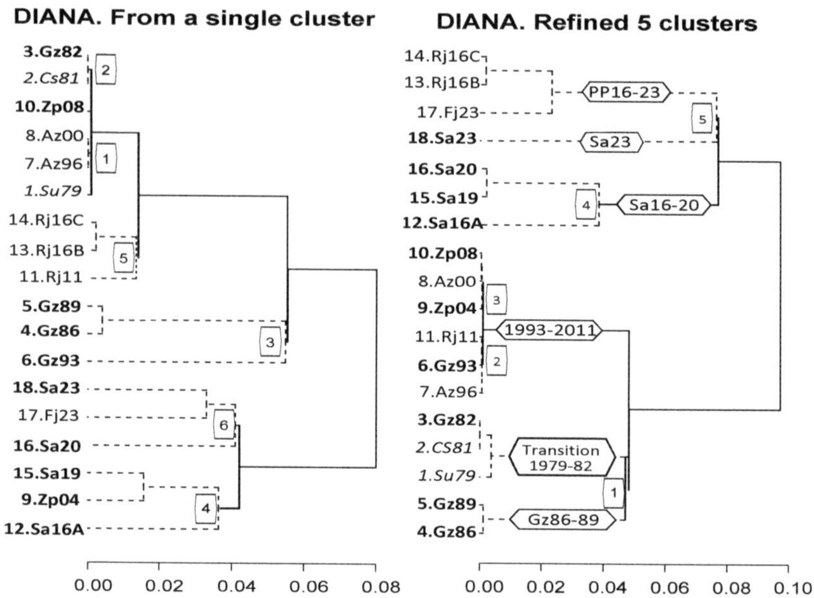

Figure 4.5: Dendrograms DIANA with size constraint equal to three (left) and refined (right).

In some cases, users may opt to initiate the process from a particular configuration, knowing the number of clusters and which points belong to which cluster, while maintaining the size constraint. As previously mentioned, the criteria employed might have led to the formation of clusters containing a large number of points. If partitions are required to reduce the size while maintaining the restrictions, it is possible to create a hierarchical cluster from this configuration. This consolidation or refinement process can help to achieve this objective and improve understanding of how agglomeration occurs within the initial clusters.

The hierarchical_clustering function from the **ssclust** package will be used again to implement this idea. For instance, we will start with the six clusters obtained from DIANA (Table 4.4, sixth row, DivS), which employed a minimum size of three points. The next step in the process is to consolidate the elements of each cluster. The results are shown in the last row of Table 4.4, (labeled Div3). Five clusters are determined in this case by the algorithm, which can be compared with the results of the configuration of five clusters without restriction (WCns5).

The ARI value (0.33) indicates low similarity between the constrained and unconstrained configurations, which contrasts sharply with the high values of cophenetic correlation (0.96) and Baker's coefficient (0.96). To understand this difference, the corresponding dendrogram was constructed in Figure 4.5. It reveals the same two major periods as in the unconstrained case (1979-2011 at the bottom of the dendrogram and 2016-2023 at the top). The structural difference is small: the Transition 1979-1982 period is now joined to Gz86-89 before the 1999-2011 period, which alters the agglomeration order observed in the unconstrained configuration.

In conclusion for this section on size restrictions, the introduction of a small increase in the minimum size from two speeches to three is evident in the significant changes to the results using DIANA, which proposes six clusters. The use of consolidation or refinement, using a hierarchical cluster based on the DIANA solution, has reduced the number of proposed clusters to five in this case. This refinement confirms two key points:

1. Although the direct comparison between the constrained and unconstrained clusters does not align under the ARI criterion, they exhibit a high degree of similarity based on the correlation coefficients.
2. The resulting dendrogram is much closer to the unconstrained case than the initial DIANA tree.

Therefore, the use of this consolidation is highly recommended.

4.7 Clustering after detrended CA

As previously mentioned, the Guttman effect frequently emerges when one of the variables, in this case the documents, exhibits an order relationship between the modalities, as illustrated in Figure 4.1. The symmetry of the functions of rows and columns in CA and the distributive equivalence property imply that this effect for the documents also occurs in the first factorial plane for words (see Figure 4.2, left). The script **chap4.e** contains instructions necessary to obtain the results shown in this section.

The second axis can be related to a quadratic function of the first axis, weighted by the length of the documents. This results in an adjusted R^2 of 0.43 after eliminating the first parameter because it was not significant. If the first axis is folded at zero, and only absolute values are considered for the second axis, a lower value of 0.35 is obtained for the fit (adjusted R^2). In this instance, the third axis does not demonstrate a cubic relationship with the first, as the adjusted R^2 is only 0.08.

Hill and Gauch (1980) also point out that, as a result, the positions of the elements tend to be more compressed at the ends of the axes than in the central part. They maintain that this structure is artificial and complicates interpretation. In the case of the example analyzed, the horseshoe effect is evident, but it is not clear that the highest density of speeches or words occurs at the extremes of the axis.

The *detrended correspondence analysis* (DCA) method was developed in the field of ecology and vegetation science to address these two challenges: flattening the arc and rescaling the positions of rows or columns along an axis (gradient). It is implemented in the decorana function from the **vegan** package (Oksanen and Minchin, 1997). DCA is initiated from the results obtained from a CA.

To detrend the first axis, the decorana function divides its coordinates into segments, with 26 being the default. Please note that the results are sensitive to this parameter. The eigenvalues decrease as the number of segments increases. Each segment is treated independently of the others and rescaled so that the average of the second-axis coordinates for each segment is zero. This process involves straightening the curve to remove the arc effect, with the objective of ensuring that the points along the first axis are dispersed at a constant rate. Finally, an iterative process is employed to refine the solution, with a focus on shifting the edges between segments to enhance robustness and prevent excessive reliance on the initial line separation.

As shown in Table 4.5, the results obtained for a DCA with 26 segments and rescaling of the axes with four iterations are presented. The function vegan::decorana only displays four axes. The DCA eigenvalues (first row) are closely aligned with the CA eigenvalues (last row), with the first factor having a significantly higher value than the rest.

Additive eigenvalues consider the effects of removing the prior axes; they can be assumed to add up to the total inertia (scaled chi-square equal to 1.0776). A comparison of the two sets of eigenvalues reveals only slight differences, suggesting a weak correlation between the four DCA axes, which is consistent with the intended goal of detrending.

Table 4.5: DCA and CA eigenvalues.

DCA	DCA1	DCA2	DCA3	DCA4
Eigenvalues	0.1732	0.0767	0.0626	0.0497
Additive eigenvalues	0.1732	0.0767	0.0625	0.0496
Axis lengths	1.4797	1.41105	1.1363	1.0393
CA eigenvalues	0.1761	0.0938	0.0891	0.0820

The length of the first DCA axis is interpreted in standard deviation (SD) units. This measure is also known as compositional dissimilarity or beta diversity in the ecology field of ecology. It is used to assess the homogeneity or heterogeneity of the dataset. It quantifies the amount of change (turnover) across the speeches. If all DCA axis lengths are less than 1.5, it suggests that speech topics renew very slowly along the gradient. As shown in the fourth line (axis lengths) of Table 4.5, this condition is met, indicating that the composition of speeches remained relatively stable over time.

To determine whether performing a cluster analysis based on the DCA would significantly improve the classification once the data has been detrended and the arc removed, it is necessary to analyze the coordinates obtained from the DCA and compare them with the CA coordinates. As illustrated in Table 4.6, the weighted linear correlations between CA coordinates (Dim) and DCA coordinates are displayed, with the limitations that only four axes are provided by decorana. The first axis remains practically unchanged when performing the DCA, except for a reversal of the sign. This is confirmed by computing the weighted Pearson correlation with the chronology for each DCA axis (−0.950, 0.112, −0.200, 0.063), demonstrating that the DCA has not eliminated the chronological component, which is essential for understanding the data.

The second CA axis has a significant correlation of 0.674 with the second DCA axis. The remaining correlations are not significant, except for axis 4 CA with axis 2 DCA.

Table 4.6: Pearson weighted correlations between CA (Dim) and DCA coordinates.

CA	DCA1	DCA2	DCA3	DCA4
Dim 1	−0.999***	0.061	0.062	−0.029
Dim 2	−0.081	0.674***	−0.152	−0.079
Dim 3	−0.112	−0.179	−0.196	−0.127
Dim 4	−0.004	−0.615***	−0.269	−0.020
Dim 5	−0.050	−0.209	0.409	−0.107

Significance codes: * $p < 0.10$, ** $p < 0.05$, *** $p < 0.01$

In order to verify the effects of the changes in the first factorial plane and relate them to Figure 4.3, Figure 4.6 shows the DCA results in the first factorial plane. The contrast between discourse 1.Su79 and the discourses of González (4.Gz86, 5.Gz89, and 6.Gz93) remains on the second axis. However, the coordinates of the other discourses on this axis have been flattened, which complicates interpretation.

The first axis indicates a considerable temporary shift for Rajoy's three speeches (11.Rj11, 12.Rj16B, and 14.Rj16C), suggesting a potential deviation from the consistent progression of the issues.

The procrustes function compares two configurations by rotating one into alignment with the other to minimize the sum of squared differences. Although it is possible to compare the shapes based on their differing number of axes (five for CA and four for DCA), we chose to compare them by using only the first four CA factors to match the dimensionality of the DCA results. This will determine if there are any differences in the cluster analysis results using CA and DCA.

As illustrated in Table 4.7, the results clearly indicate that, with the exception of the reflection indicated by the negative sign, the first axis is practically the same when using CA and DCA (correlation: −0.995). This strong relationship indicates that chronology is the dominant explanatory factor. A similar strong correlation is observed between CA axis four and DCA axis three (−0.935). This swapping was not observed in the non-rotated

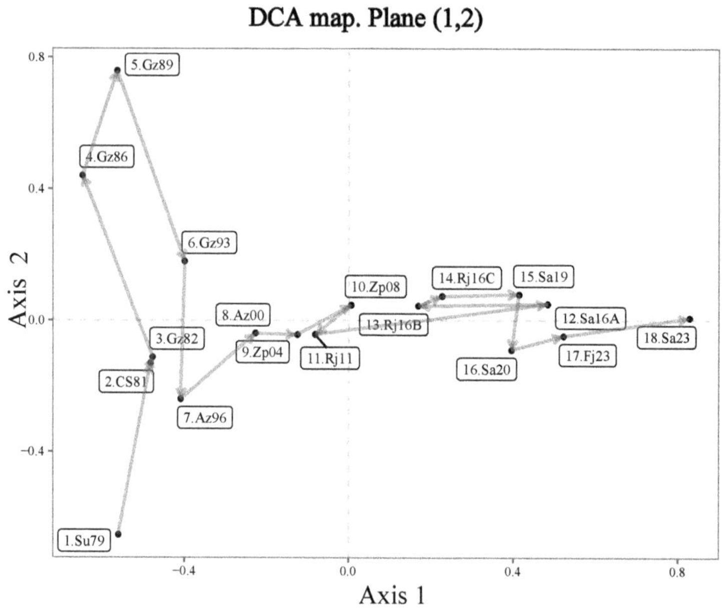

Figure 4.6: Representation of documents on the first DCA plane.

configurations. The second axis displays a lower correlation (0.790). The closeness between the eigenvalues of factors three and four suggests that CA axis three is also swapped with DCA axis four (−0.738). Additionally, it has come to our attention that CA factor two now partially aligns with DCA factor four (−0.574), and CA factor three partially aligns with DCA factor two (−0.611).

Table 4.7: Procrustes matrix between CA (Dim) and DCA coordinates.

CA	DCA1	DCA2	DCA3	DCA4
Dim 1	−0.995	0.046	−0.077	−0.044
Dim 2	0.005	0.790	−0.215	−0.574
Dim 3	0.082	−0.611	−0.274	−0.738
Dim 4	0.057	0.001	−0.935	0.352

In our case, obtaining different configurations, including rotated ones, initially leads us to wonder whether DCA facilitates interpretation in terms of coordinates. To this end, Figure 4.7 shows the dendrograms obtained from CA (on the left) and DCA (on the right) based on the coordinates of the first four dimensions.

By reducing the number of dimensions to 4, the three Sánchez speeches are now grouped together (15.Sa19, 16.Sa20, and 18.Sa23), while 12.Sa16A is grouped with the three PP speeches.

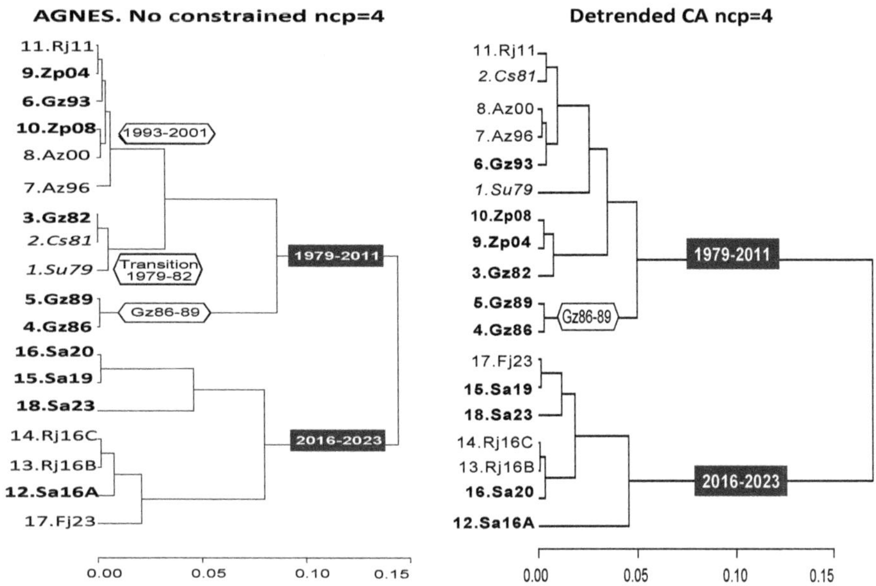

Figure 4.7: Dendrograms from CA (left) and DCA (right) from four dimensions.

The DCA dendrogram (Figure 4.7, right) shows a notable distinction from the CA tree, as it disrupts the previously delineated periods of discourse. Despite the overall high correlation coefficients (0.92 for cophenetic and 0.94 for Baker's coefficient), the classification of the speeches only coincides for the two- and three-group cut-offs, both yielding an ARI coefficient of 1.

When the dendrograms are divided into four distinct groups, the ARI index is 0.82, dropping to 0.45 for five clusters and only 0.31 for six clusters. In summary, the results obtained with DCA differ considerably from those obtained with CA, both in terms of coordinates and dendrograms. Interpretation has become more complex, both from the point of view of chronology and of the politicians who delivered the speeches. In this particular instance, CA has been demonstrated to be a more effective approach than DCA. This finding aligns with numerous other studies in this field.

As stated in the works of Økland (1990) and McCune and Grace (2002, chapter 20), a comprehensive analysis of the shortcomings and advantages in this sense had already been presented. One criticism of DCA is that it often creates artificial structures that hinder the analysis of data structure, which has occurred in our study. Other researchers argue that detrending does not solve the problem of compression of the points at the ends of the axes. They also point to the instability that results from changing the number of segments. A review of this criticism is available in Oksanen and Minchin

(1997), Legendre and Legendre (2012, pp.471-472), or Borcard et al. (2018, p.139).

4.8 Canonical correspondence analysis CCA

Canonical correspondence analysis (CCA) is a type of constrained analysis that can be applied jointly with the DTM matrix and a set of variables to extract the relationships and uncover the multidimensional structure of the DTM matrix.

In its original formulation, ter Braak (1986) described CCA as an extension of CA, incorporating external information known about the contingency table to be analyzed. These variables, often referred to as environmental variables, can be quantitative (Lebreton et al., 1988a) or nominal (Lebreton et al., 1988b) and are applied as a CA restriction. This restriction allows us to better explain some DTM patterns using these environmental variables, such as spatial or temporal factors, or individual characteristics gathered from open-ended questionnaire responses.

For this reason, CCA is a good choice when we are not interested in the overall structure of the DTM matrix, but rather in analyzing the part related to environmental variables. It is easy to compute the difference between the original DTM and the DTM that can be reconstructed from the CCA results. In this way, it is not recommended to perform CCA with all environmental variables, but rather, only those that have been proven to explain something of the overall structure should be used.

Oksanen considers that CCA is able to correct the Guttman curve found in the first plane of CA by imposing linear constraints on the configuration. However, he points out that the curve can only be avoided by imposing a small number of restrictions that are not related to each other in a curvilinear fashion. The curve may reappear even with two poorly chosen restrictions or a single experimental variable, as in the case we present, which only contains the chronology.

Let $\mathbf{P}$ be the DTM matrix divided by the grand total, such that p_{ij} are the cell proportions. The standardized residual y_{ij} is the deviation of the observed proportions p_{ij} from independence, where $p_{i\cdot}$ are the marginals of the documents and $p_{\cdot j}$ the marginals of the words:

$$y_{ij} = \frac{p_{ij} - p_{i\cdot} \cdot p_{\cdot j}}{\sqrt{p_{i\cdot} \cdot p_{\cdot j}}}.$$

CCA is an asymmetric method that, by denoting the response matrix as $\mathbf{Y}$, employs a weighted regression to predict $\mathbf{Y}$ from an $\mathbf{X}$ matrix containing explanatory environmental variables. Matrix $\mathbf{X}$ is centered and must be standardized if the variables are in different units. The data in the regression analysis are weighted by the row marginal proportions of the documents, $p_{i\cdot}$. Let $\mathbf{D}_{p_{i\cdot}}$ be the diagonal matrix containing these weights.

The fitted component of $\mathbf{Y}$, denoted as $\hat{\mathbf{Y}}$, can be calculated as follows:

$$\hat{\mathbf{Y}} = \sqrt{\mathbf{D}_{p_{i\cdot}}}\mathbf{X}\left[\mathbf{X}^{\mathbf{T}}\mathbf{D}_{p_{i\cdot}}\mathbf{X}\right]^{-1}\mathbf{X}^{\mathbf{T}}\sqrt{\mathbf{D}_{p_{i\cdot}}}\mathbf{Y}.$$

The residual matrix $\mathbf{Y}_{\mathbf{res}}$, representing the unexplained variation, is calculated as the difference between the observed and fitted components:

$$\mathbf{Y}_{\mathbf{res}} = \mathbf{Y} - \hat{\mathbf{Y}}.$$

Finally, the CCA solution is found by performing an eigenvalue decomposition on the cross-product matrix, $\hat{\mathbf{Y}}^{\mathbf{T}}\hat{\mathbf{Y}}$.

The script in this section, labeled **chap4.f**, contains the **R** instructions required to generate the displayed results. The **vegan** package will be used again, this time with the cca command, which is based on the algorithm described in Legendre and Legendre (2012).

Matrix $\mathbf{Y}$ in the cca function is the DTM matrix, and $\mathbf{X}$ is the chronology variable. While there are several ways to express this relationship, the most flexible method is to use a formula. In our case, this formula uses the DTM matrix obtained from the TextData function of the **Xplortext** package:

```
vegan::cca(as.matrix(TD$DocTerm)~TD$SourceTerm.quant$chronology)
```

The total inertia, using chi-square distance, computed previously in this chapter, is (1.0776). The constrained inertia (explained variance) is 0.1643, leaving the unconstrained inertia (unexplained variance) as the difference, 0.9134. In percentage terms, the chronological variable $\mathbf{X}$ explains 15.24% of the structure of matrix $\mathbf{Y}$. When corrected using a bootstrap procedure, the adjusted R^2 value for this explanation is 9.71%.

When selecting variables for the $\mathbf{X}$ matrix, it's important to ensure they aren't multicollinear (or highly correlated). If they are, both the R^2 and adjusted R^2 values can be artificially inflated considerably. Indices such as Tolerance, Variance Inflation Factor (VIF), or the Condition Index can be used to check for multicollinearity.

Let I=18 be the number of documents, J=1941 be the number of different words, and M=1 be the number of variables in the $\mathbf{X}$ matrix. In this case, the maximum number of canonical axes in CCA is limited to one, determined by the minimum value of $(I-1)$ and M.

The difference between the original $\mathbf{Y}$ matrix and the fitted values obtained using the $\mathbf{X}$ matrix is the residual matrix. The unconstrained eigenvalues (CA1, CA2, etc.) result from performing a CA on this residual matrix. In this case, the maximum number of unconstrained axes is 16, calculated as:

$$\min[(I-1),(J-1)] - M = \min[(18-1),(1941-1)] - 1 = 16.$$

Before continuing with the analysis, it is advisable to check the significance of this relationship. The F-test (with a statistic of 2.88 and calculated using χ^2 values) yields a p-value of 0.001. Since this p-value is less than the significance level of $\alpha = 0.05$, we reject the null hypothesis that the relationship between matrices **Y** and **X** is random.

To compare the results of the CCA with those previously obtained with CA, Table 4.8 shows the correlations between the coordinates of the first factors for the documents (or rows). Similarly, the correlations between the coordinates for the words (or columns) could have been calculated.

Table 4.8: Pearson correlations between CCA and CA. Eigenvalues and explained percentage.

	Correspondence Analysis CA						
CCA	Dim1	Dim2	Dim3	Dim4	Dim5	Eigen	Expl.Pct.
CCA1	−0.991	−0.053	−0.135	−0.056	−0.132	0.164	0.152
CA1	0.036	1	0.014	−0.016	0.078	0.094	0.087
CA2	−0.011	−0.022	0.970	0.232	0.177	0.090	0.083
CA3	−0.236	−0.044	−0.202	0.905	0.327	0.085	0.079
CA4	0.152	−0.045	0.027	0.282	−0.853	0.074	0.069
CA5	0.076	0.062	0.017	0.161	0.382	0.064	0.059
Eigenvalue	0.176	0.094	0.089	0.082	0.070		
ExplainedPct.	0.163	0.087	0.083	0.076	0.065		

The row labeled CCA1 shows the correlation between the coordinates of the single constrained axis and the coordinates of the original unconstrained axes from the initial CA. The canonical axis CCA1 is essentially aligned with the original Dim1 from CA because they are highly correlated ($r = -0.991$), noting that the negative sign is arbitrary and only indicates an axis reflection. This indicates that CCA1 contains the chronological component that captured the first dimension of CA.

The correlation table demonstrates that the second unconstrained axis of the original CA becomes the primary axis of the residual variation (as shown by the perfect correlation between CA1 and Dim2). Analyzing the subsequent axes, the strong diagonal values (close to 1) indicate that the remaining residual axes correspond almost directly to the remaining original CA dimensions: CA2 correlates highly with Dim3 ($r = 0.970$), CA3 correlates highly with Dim4 ($r = 0.905$), and CA4 correlates highly with Dim5 ($r = -0.853$).

The significantly lower correlation values observed between the canonical axis CCA1 and the subsequent unconstrained CA dimensions Dim2 through Dim5 can be explained by analyzing Table 4.1. The table shows that the chronology variable's square cosine (cos^2) value—indicating its contribution to each axis—is overwhelmingly concentrated on Dim1 (89.28%), leaving negligible contributions to the remaining dimensions: 0.01% (Dim2), 0.22% (Dim3), 1.96% (Dim4), and 3.97% (Dim5), respectively.

This confirms that the chronological effect has been extracted and captured by the canonical axis CCA1. The remaining variation was removed from the unconstrained analysis, which is why the unconstrained residual axes (CA1, CA2, etc., performed on $\mathbf{Y_{res}}$) are perfectly correlated with the original Dim2, Dim3, and so on. Furthermore, it should be noted that the construction of each subsequent axis is conditionally independent of the previous axes due to the orthogonality restrictions inherent in the ordination method.

In relation to the interpretation of eigenvalues, in general terms, those obtained for CCA will be lower than those obtained for CA, because part of the CA inertia is now explained by the environmental variables. The stronger the relationship between a CA dimension and environmental variables, the greater the reduction in eigenvalues. However, it should be noted that the factor structure may have changed. In our case, the CCA eigenvalue (0.164 CCA1) accounts for 15.2% of the total inertia (1.0776), so once this amount is removed, the remaining unexplained or unconstrained part (0.9134) must be distributed among fewer dimensions (16 compared to 17 in CA).

Constructing a factorial plane with CCA1 (the canonical axis) and CA1 (the first unconstrained residual axis) yields a figure that is practically the same as the original first factorial plane formed by Dim1 and Dim2 in Figure 4.1 (first row, left). This result suggests that the CCA process did not successfully eliminate the Guttman effect from the configuration. For this reason, we will not proceed to graphically represent the speeches or perform the corresponding cluster analysis in this section.

To identify words with a temporal relationship, we analyze their positions using only the CCA1 coordinates. Initially, dividing the 1,941 words into two clusters (1,034 words and 907 words) explains 62.8% of the variance (as shown by the solution's explained inertia in Figure 4.8). The elbow plot further indicates that increasing the solution to three clusters increases the explained variance to 80.01%.

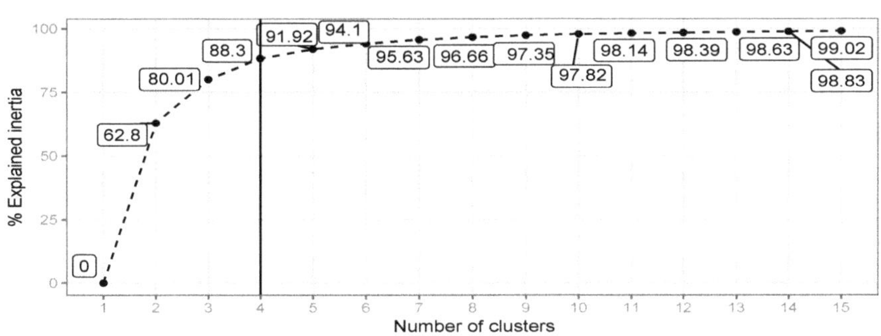

Figure 4.8: Elbow representation for cluster of words from CCA1 coordinates.

4.9 Contiguity-constrained hierarchical clustering

As previously mentioned, a corpus can be restricted in a number of ways to better understand its structure. One example of a constrained relationship is contiguity, which can be a spatial or temporal restriction (Chavent et al. (2015); Chavent et al. (2018)). In this context, individuals can only join if they are neighbors of others, either through spatial or temporal contiguity. We will use the term *contiguity* in a broad sense. The **ClustGeo** package (Chavent et al., 2018) facilitates the implementation of hierarchical clustering methods incorporating spatial constraints. The method utilizes two dissimilarity matrices:

1. $\mathbf{D_0}$, a matrix containing the dissimilarities computed in the *feature space* (e.g., socio-economic variables).
2. $\mathbf{D_1}$, a matrix created by computing the dissimilarities in the constraint space.

The values in $\mathbf{D_1}$ can be based on geographical distances between each pair of points. Alternatively, $\mathbf{D_1}$ can be constructed from a dichotomous relationship (an adjacency matrix) where 1 indicates a connection between two points is possible and 0 indicates no connection is possible.

However, this restriction can be overly restrictive in certain situations. The presence of a document that significantly deviates from both previous and subsequent documents can disrupt the cohesion of two groups that bear a strong resemblance to each other. This document is considered a singleton, which raises the question of whether it should be initially removed from the study when performing cluster analysis. The algorithm can be further refined by incorporating an alpha parameter that ranges from 0, indicating that only adjacent individuals or clusters can merge, to 1, denoting the absence of contiguity restrictions. As we will discuss later, the implementation of this weighting mechanism depends on the aggregation method selected.

Often, the corpus consists of a series of ordered documents, arranged by production date as in the case of the inaugural speeches we are using in this chapter. Including the contiguity constraint means that documents or clusters in nodes in the hierarchy can only be aggregated if they are contiguous. Among the possible applications, even a document can be split into parts, considering each one as consecutive in a temporal order, as the dialogue of a movie. In other fields, such as ecology or epidemiology, restrictions can be established both in time and space, without forgetting that the primary objective is to group similar individuals together.

In some cases, the objective may be to measure the association between vocabulary or words and qualitative or quantitative variables trying to determine the specific topics in each category. Chapter 1 of this book introduces the concept of the ALT, which is created by crossing the DTM matrix with one qualitative variable. The categories into which documents are aggregated are collapsed into a single row, while the columns (words) remain unchanged. The ALT table has as many rows as there are categories of the qualitative variable, and as many columns as there are words in the vocabulary. For instance, in the case of open-ended questions, the age of the respondents can be analyzed to identify age groups with similar lexical profiles and to detect breaks.

4.9.1 Chronological contiguity

In this chapter, the example of the Spanish speeches has been used, applying various constrained methods and verifying the high correlation of the results with the chronology in the first axis of the CA for the documents. However, it is difficult to measure the importance of chronology on a corpus structure because this chronological relationship can be divided between different factorial axes, or it may not even be significant in the structure derived from CA. As was previously discussed, HCA on the documents can use the coordinates provided by CA as input data, bearing in mind that these coordinates are in a Euclidean space. For a chronological corpus, the time-adjacent texts are expected to have more similar vocabularies than the long-interval separated texts. If the differences among the clusters correspond only to thematic variations over time, the nodes of the dendrogram created using HCA will only group contiguous elements.

To account for the effect of chronological restrictions, we can use *hierarchical contiguity-constrained clustering methods* (HCCC) (Lebart (1978); Legendre and Legendre (2012); Borcard et al. (2018); Gançarski et al. (2020)), taking the coordinates from a CA as input data.

In general, dendrograms constructed from HCA and HCCC are distinct due to the evolution of vocabulary over time. The phenomenon of linguistic variation is characterized by the appearance of certain words at specific temporal points, while others undergo extinction, and yet others demonstrate a capacity for longevity. The discernment of substantial discrepancies between HCA and HCCC is of paramount importance, as it furnishes critical insights into the underlying knowledge of the structure of the corpora.

A variety of modifications to classic clustering algorithms have been documented in the literature to introduce this type of restriction (see, for example, Murtagh (1985); Legendre and Legendre (2012); Bécue-Bertaut et al. (2014a); Bécue-Bertaut (2019)).

4.9.2 Reversals in HCCC

One of the decisions that must be made when performing a cluster analysis is how to select the aggregation method. Lance and Williams demonstrated that seven HCA strategies of aggregation can be related using a linear formula with four parameters (Lance and Williams, 1967), and Murtagh and

Contreras (Murtagh and Contreras, 2012) extended this idea to hierarchical *self-organizing maps* (SOM), and *grid-based clustering.*

Ferligoj and Batagelj (1982) discovered that imposing a restriction on HCA can result in the appearance of inversions (also called reversals or crossovers) in the dendrogram. A reversal is a departure from ultrametricity, meaning the merging dissimilarity of the two clusters is lower than the dissimilarity of a previous merge. The result is typically displayed in the dendrogram as a crossover. For examples, see Owsiński et al. (2021). One solution is to use aggregation methods that are guaranteed not to produce reversals, such as single linkage (also known as nearest neighbor) or complete linkage (also known as furthest neighbor) (Glasbey, 1987).

4.9.3 Complete linkage method

Next, the chronology restriction will be included in the cluster analysis using an ascending hierarchical iterative procedure and the complete linkage method. The LexCHCca function from the **Xplortext** package (Bécue-Bertaut et al., 2024) is used for the implementation of this chronological contiguity Section.

1. A distance matrix $\mathbf{D}$ between the terminal nodes is constructed by computing the Euclidean distance $d(i,j)$ between each pair of documents.
2. A similarity matrix $\mathbf{S}$ with elements $s(i,j)$ is computed from the distance matrix. The $s(i,j)$th entry of this matrix is equal to $d_{max} - d(i,j)$, where d_{max} is the maximum distance between all pairs of terminal nodes adding a small epsilon value.
3. The contiguity $\mathbf{C}$ matrix is built using binary values. The value is 1 if points i and j can be joined, and 0 otherwise.
4. The Hadamard product $\odot$, which computes the element-wise product, is used to apply the contiguity constraint to both the distance and similarity matrices:

$$\mathbf{D}^* = \mathbf{D} \odot \mathbf{C}$$

$$\mathbf{S}^* = \mathbf{S} \odot \mathbf{C}$$

5. The maximum value of $\mathbf{S}^*$ is sought by finding the indices i and j (where $i < j$) for which this maximum is achieved. The corresponding nodes are then aggregated to form a new cluster, which replaces node i.
6. The similarity values between this new cluster and all remaining clusters are then calculated. For each cluster k remaining in the matrix, the new similarity is calculated using the complete linkage (furthest neighbor) rule. For each row of $\mathbf{S}^*$, the minimum

value of columns i and j is calculated and stored in column i. The diagonal for the new similarity value is changed: $s^*(i,i) = max\{s^*(i,i), s^*(j,j)\}$. The similarity matrix is updated, filling in the data corresponding to the new group in row i and column i, while column j and row j are deleted. The similarity values between the rest of the clusters remain constant. A similar process is followed for the distance matrix, and the results are saved to build the final dendrogram.

7. The contiguity **C** matrix must be changed to adjust the values to the left and right of i and j in order to establish the new contiguity values (1 for contiguity and 0 for non-contiguity).
8. While all documents are not found in the same node, return to step 4, else: stop.

The **Xplortext** package employs the same criteria outlined in the **FactoMineR** package, focusing on ratios between inertia values, to suggest a number of clusters, as explained in Chapter 3. It suggests that there are six groups, as shown in Figure 4.9 (on the first line on the left, labeled *Chronological constrained. No test*) (script in **chap4.g**). It has been established in this example that chronology constitutes a fundamental component of the data structure. The results obtained through the application of temporal contiguity using the complete clustering method are consistent with the findings reported in Section 4.4 using the Ward method.

The first period consists of speeches given between 1979 and 2011. The small distances between groups in Ward's method are now more pronounced when applying the contiguity restriction. The first three sub-periods coincide with those obtained previously.

- *1.Su79 – 3.Gz82.* This first sub-period corresponds to the three speeches concerning the Spanish Transition, dating from 1979 to 1982.

- *4.Gz86 – 5.Gz89.* The second sub-period comprises two speeches delivered by the same author, Felipe González (4.Gz86 and 5.Gz89). These speeches exhibit distinctive elements of his rhetorical style and suggest a disruption in the sequence of events, marked by reforms and changes.

- *6.Gz93 – 11.Rj11.* The third sub-period includes six speeches, covering approximately 18 years, and aligns with the cluster that does not have any chronological restrictions.

  Without the chronological restriction, this sub-period merged with the first sub-period (which corresponds to the Transition), a result attributable to the significant differences between González's two central discourses (4.Gz86 and 5.Gz89) and the rest.

  Because chronology makes it impossible to group Cluster 1 (1.Su79 – 3.Gz82) and Cluster 3 (6.Gz93 – 11.Rj11), the three subgroups remain, although the

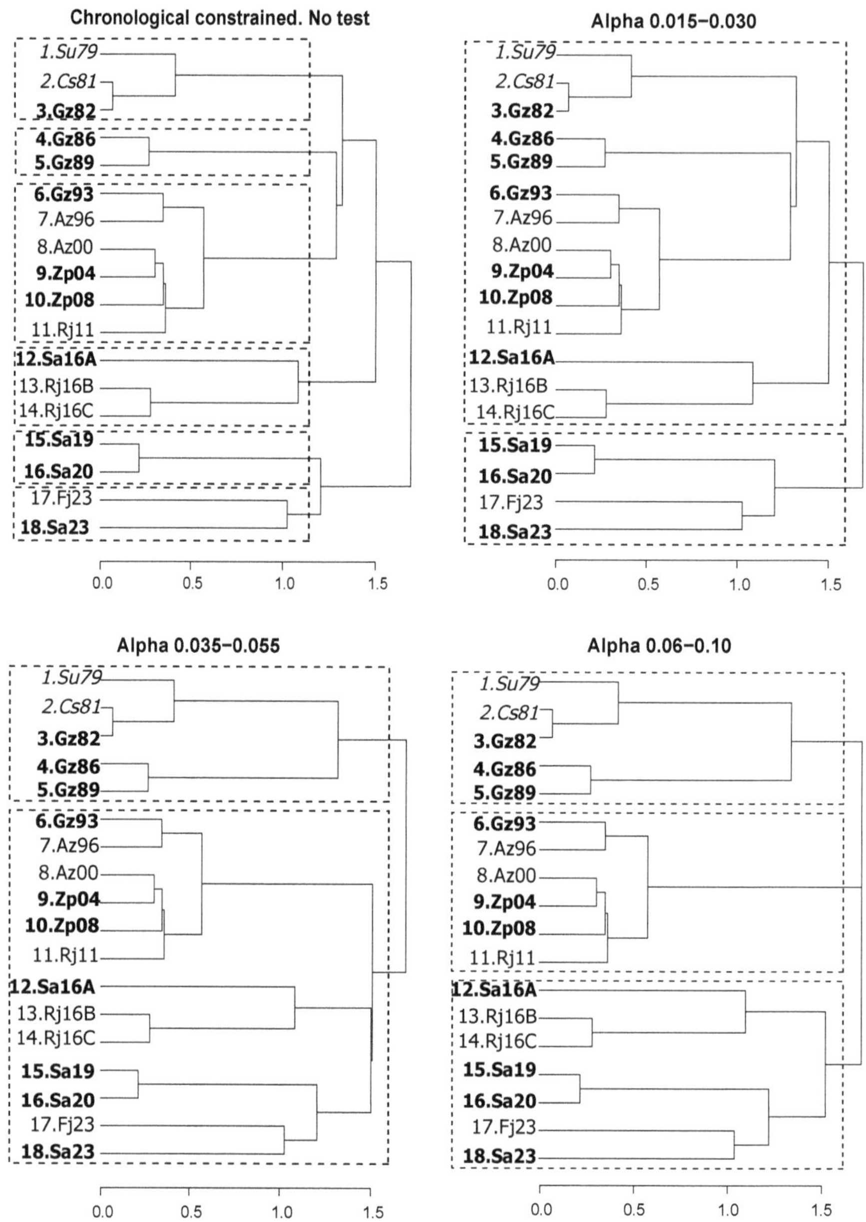

Figure 4.9: Dendrograms with chronological constraint and several values of alpha in the stopping rule.

second and third clusters merge first. However, the distances at which these three groups merge are very similar in both cases.

The second period, initially without a chronological constraint, covered the years 2016-2023. Its structure has changed considerably since then to include a chronological restriction. To understand the departure from the previous period and the differences between the two, it is important to note that on December 20, 2015, general elections were held under Rajoy's leadership, resulting in seats in Parliament for 19 different political parties.

In a highly polarized context, on January 23, 2016, King Felipe VI proposed Mariano Rajoy as the candidate, but Rajoy declined the offer in the face of certain defeat. Specifically, the vote projection was 123 votes in favor, 186 against, and 41 deputies who did not indicate their position.

In a second attempt, the King proposed Pedro Sánchez (12.Sa16A). He failed to obtain a majority in the second vote, held on March 4, receiving 131 votes in favor and 219 against. Finally, the deadline expired on May 2, 2016, and new elections were called, which were subsequently held on June 26, 2016.

The PP won the elections again, though the overall distribution of seats between the left and the right remained constant, with a total of 17 different political parties winning seats. The King nominated Mariano Rajoy, as the leader of the party with the most deputies, to form a government. The first investiture session for the PP candidate was scheduled for August 30, 31, and September 2, 2016. The candidate was rejected with 170 votes in favor and 180 against.

In the absence of viable candidates, some PSOE members negotiated their congressmen's abstention from a new vote. This created a crisis within the party, resulting in the resignation of 17 members of the PSOE's Federal Executive Committee and the General Secretary, Pedro Sánchez, on October 1. This strategy enabled Mariano Rajoy to be elected on October 29, 2016, with 170 votes in favor, 111 against, and 68 abstentions from the PSOE. This speech (14.Rj16C) marked the conclusion of a period defined by the norms of parliamentary courtesy. Since that moment, polarization in the Congress of Deputies has been very strong.

Recalling the clusters obtained for the period from 2016 to 2023 Rajoy's two previous speeches were grouped with the last speech by the PP candidate (17.Fj23). Additionally, Sánchez's three speeches (12.Sa16A, 15.Sa19, and 16.Sa20) were grouped together, leaving Sánchez's last speech (18.Sa23) separate. The introduction of the contiguity restriction (Figure 4.3, left) alters the groupings for this period. Specifically, Sánchez's two speeches (15.Sa19 and 16.Sa20) now form a group, which then merges into a final group with (17.Fj23 and 18.Sa23).

To test the similarity between the classification results without and with contiguity constraints, the Adjusted Rand Index (ARI) was calculated, yielding a high value of 0.83. This high agreement suggests that a significant degree of chronological restriction was already inherent in the original data structure, thereby reaffirming the importance of chronology in this study. Similar findings were obtained using the cophenetic correlation (0.80) and Baker's gamma correlation (0.71).

4.9.4 Local stopping rules in contiguity constrained clustering

The decision on the number of clusters selected in a cluster analysis aims to represent a structure similar to that which exists in reality. Non-hierarchical procedures generally require the user to specify this parameter in advance, before performing any grouping. In this case, the choice of the number of clusters is usually made by comparing the results for different values of the number of clusters. Hierarchical methods delay this decision until the aggregation process is complete, for example by analyzing the dendrogram. The implementation of a stopping rule in a cluster analysis is intended to automatically provide a result for this decision. The process incorporates specific criteria, which will be evaluated prior to the presentation of the final results and condition them.

During this decision-making process, there are two errors that could be made. The first error occurs when the stopping rule indicates a number of clusters greater than those existing in the true structure. This can complicate the characterization of each cluster, making it appear fictitious and artificial. The second error occurs when the number of clusters chosen is less than the actual number. The result is excessive heterogeneity among individuals in the same group and insufficient separation between the clusters obtained. The balance to be sought will depend on the context in which the analysis is carried out.

As previously discussed, HCCC possesses the capacity to group contiguous speeches; however, it is also possible that the resulting nodes may group excessively heterogeneous clusters, in other words, building artificial clusters. From a textual point of view, it is important that a cluster is interpretable. To achieve this, the individuals that make up the cluster must have similar lexical profiles and clusters with different lexical profiles should not be joined.

Local stopping rules have been applied in cluster analysis with several objectives (Milligan and Cooper (1985); Gordon (1999); Arbelaitz et al. (2013)). In many cases, the aim is to determine the most appropriate number of clusters by incorporating a certain degree of objectivity into the decision. Furthermore, it is essential to evaluate the stability of the results to ensure the reliability and validity of the study's findings, so that even minor alterations in the sample do not result in significant changes in the composition of the clusters.

According to Milligan (1985), the search for an optimal rule compared to others requires the assessment of procedures that are method independent. Nevertheless, this is not always possible, and sometimes, the stopping rules can only be applied to certain methods. Therefore, the grouping system used is important. However, the results of these comparative studies should be interpreted with caution, because they have often been carried out using artificial data whose structure has been defined in advance.

Another difficulty that may be encountered is that, similar to restrictions, stopping rules can lead to inversions. As previously stated, in order to avoid the occurrence of inversions when employing contiguity-constrained aggregation methods, it is advisable to utilize complete linkage aggregation (Nielsen, 2016).

In order to build clusters that accurately correspond to the data structure and avoid building clusters whose elements are heterogeneous, we will implement the solution proposed by Legendre and Legendre (2012), which was originally designed to detect discontinuities in multi-species time series (Legendre et al., 1985). The number of clusters is determined automatically, but it depends on an alpha parameter that must be specified beforehand.

The iterative procedure for building the contiguity cluster must be modified to include a statistical test prior to authorizing the fusion of two terminal or intermediate nodes. The first change is related to point 5. The chronology restriction will be included in the cluster analysis using an ascending hierarchical iterative procedure. The LexCHCca function will be used again to apply the stopping rule.

5.a The first step in this section remains the same, corresponding to the first part of step 5 of the analysis without restriction. It involves finding the maximum value of the similarity matrix $\mathbf{S}^*$ and retrieving the i row and j column corresponding to this maximum.

5.b If each of the two clusters contains only one individual, the test is not applied, and elements i and j are merged. In this case, skip the remaining steps in step 5 below and proceed to old step 6.

5.c A symmetric binary matrix $\mathbf{B}$ is constructed. The first columns contain all elements belonging to the first cluster n_i, and the last columns contain all elements corresponding to the second cluster n_j. The distances between the first part of the matrix, which corresponds to the first n_i rows and n_i columns, are all zero. That is, the distance between all elements in the first cluster is equal to zero. The same procedure is followed for the part of the matrix corresponding to the last n_j rows and the last n_j columns, considering that the distances between all the elements in the second group are also equal to zero. Finally, a value of 1 is assigned to the rest of the elements in the matrix that correspond to the distance between individuals belonging to different clusters.

5.d A new binary matrix $\mathbf{B}^*$ is computed by assigning a value of 1 to values of matrix $\mathbf{S}$ greater than a cutoff point and a value of 0 otherwise. Sometimes, this cutoff point is the median. Since this may not be appropriate if the sizes of the clusters are very different, this algorithm uses the percentile of the distances according to the size of the groups.

5.e If the binary matrix $\mathbf{B}^*$ is equal to the extreme matrix $\mathbf{B}$, then we have two clusters with maximum homogeneity intra groups and maximum distance inter groups. Therefore, grouping these two clusters together is artificial and meaningless. The Mantel statistic test checks the correlation between two matrices of the same dimension. The limits and interpretation are the same as those for the Pearson linear correlation coefficient. The null hypothesis states that there is no correlation between the two binary matrices.

5.f To calculate the p-value and assess statistical significance, the rows and columns of the first matrix ($\mathbf{B}^*$) are permuted without replacement, while the second matrix ($\mathbf{B}$) is kept fixed. If the number of requested permutations exceeds the number of possible unique permutations, duplicates will not be generated.

5.g If the $\mathbf{B}^*$ matrix is equal to the $\mathbf{B}$ matrix (the most unfavorable situation for the union of the two clusters), then the Mantel test will be equal to one, and the p-value will be zero for $H_0 : r_{\mathbf{B}^*\mathbf{B}} = 0$. The null hypothesis is rejected if the p-value is less than the alpha value, and no cluster union is performed. In this case, phase 6 is not performed, and phase 7 of grouping without stopping rules must be rewritten.

If the null hypothesis is not rejected for the alpha level, the cluster union of the two elements or nodes is allowed, proceeding to old phase 6 to continue as if the test had not been performed.

7 The contiguity $\mathbf{C}$ matrix must be updated to adjust the values to the left and right of i and j in order to determine the new contiguity values (1 for contiguity and 0 for non-contiguity). If the cluster merging was performed, then rows and columns j were deleted in step 6, so matrix $\mathbf{C}$ will be adjusted to the left and right column i (and the corresponding row) by changing zeros to ones where necessary, allowing new mergers by contiguity. If clustering was not allowed, phase 6 was not performed, and the matrices remain the same size. The values corresponding to contiguous groups that were not merged are changed to zero.

Obviously, this test can be applied to non-chronological cases and with non-binary weighting that allows or disallows the union by modifying the contiguity matrix ($\mathbf{C}^*$).

The final result may only show the partial dendrograms corresponding to the elements that have been merged. However, examining the subdendrograms (the separate parts of the tree that correspond to the allowed cluster groups) is interesting because it allows one to see how documents—speeches, in this case—come together. Subsequently, the vocabulary characteristic of each period can be obtained.

To facilitate the export of the resulting tree to other **R** packages, the LexCHCca function completes the dendrogram by assigning an equal maximum distance to all groups that were not merged because of the test.

Sometimes, taking chronological order into account, it is possible to obtain elements with great disruptive power by separating two similar periods. In this case, an element acting as a singleton can be removed when building the clusters and then added to the final classification, although the dendrogram can no longer be obtained. Another difficulty that may arise when constructing a dendrogram is to find inversions or reversals, which can appear even when using the complete linkage aggregation method.

It is important to note that the dendrograms obtained with and without the stopping rule may differ, and not only because of discontinuities. Therefore, the number of clusters selected using stopping rules cannot be applied to cases without contiguity rules, as the resulting structures may differ.

In phase 5.g, it was indicated that the p-value obtained from the Mantel test is compared with an alpha value representing the significance level, which must be specified before the algorithm begins. If the p-value is less than the alpha value, the null hypothesis is rejected and the fusion of the nodes is prevented.

The clustering results change when the value of alpha changes. When alpha is small, only the sharpest discontinuities in the chronological corpus are identified, with the possibility of obtaining a very small number of clusters containing many documents each. Even if the alpha value is close to zero, no grouping may be performed.

An increase in the alpha value results in the formation of more and smaller groups, which reveals more discontinuities in the data series. It should be noted that an alpha value equal to 1 will not provide the same classification as the cluster with contiguity without stopping rules, since the stopping rule does not apply when evaluating the union of two clusters that contain only one element.

Figure 4.9 (first row, left) shows the dendrogram for the chronological constrained method with no stopping rules. Comparing this result with that obtained for alpha values between 0.015 and 0.030 (first row on the right) reveals that the two dendrograms are very similar.

In this case, the stopping rule establishes two large clusters. The first cluster contains the first 14 speeches, up to 14.Rj16C. The second cluster contains the remaining four speeches, from 15.Sa19 to 18.Sa23. This break in the dendrogram has been explained in detail in this section. As mentioned earlier, a low alpha value results in a small number of clusters. The **Xplortext** package indicates that no grouping is possible for alpha values less than 0.015.

If the alpha value is increased from 0.035 to 0.055, there are still two groups, but their composition is different. The first group contains only five speeches (1.Su79-5.Gz89). The first three correspond to the Transition period, and the last two are speeches by González (4.Gz86-5.Gz89) that have already appeared, far apart from the rest in previous analyses.

Paradoxically, the second group begins with González's last speech (6.Gz93), which divides his four speeches into two groups. This significant difference has already been mentioned above, with the latter corresponding to a speech delivered in a context in which he won the elections by maintaining a low profile despite unfavorable polls.

Increasing the alpha value to 0.06, the number of clusters goes up to three, keeping the elements of the first group and splitting the second group of speeches (6.Gz-93-18.Sa23) into two parts.

The initial cluster of this period encompasses speeches from 6.Gz91 to 11.Rj11, marked by the presence of absolute majorities or agreements among various political parties. The last period begins with Sánchez's speech in 2016 (12.Sa16A). This corresponds to the turbulent period in which Rajoy declined the King's request to form a government, which was followed by three different investiture speeches occurring in the same year. The final four speeches mark the conclusion of this period. They are grouped with the previous speeches of 2016, but there is a significant distance between them.

By increasing alpha to 0.105, four clusters are obtained, breaking the first cluster formed by the first five speeches (1.Su89-5.Gz89) into two groups that have been characterized above, corresponding to the Transition (1.Su89-3.Gz82) and to the period of change in the two central speeches by González (4.Gz86-5.Gz89) (dendrogram not included in Figure 4.9).

As has been verified, increasing the value of alpha increases the number of clusters, allowing for a more detailed analysis of the results. Obviously, the choice of an appropriate number of clusters and their structure depends on the value of alpha chosen, so comparing all possible results can be a tedious and time-consuming process. In order to facilitate this decision, it is possible to make a grid with the different results for each value of alpha and summarize them, as shown in Figure 4.10.

The variation in the different alpha values produces fairly consistent results, with only four different cluster configurations. Studying how clusters change as alpha varies also provides valuable information, helping us interpret the results and understand the structure of the data.

4.10 Conclusion

Cluster analysis, understood as an unsupervised machine learning (UML) method, is a good way to discover unknown underlying structures. Sometimes, external information about the data is available. The objective of Semi-Supervised Learning (SSL) is to combine UML with this external information. The objective is not only to obtain clusters that include this knowledge in the form of constraints, but also to enable the assignment of labels to clusters.

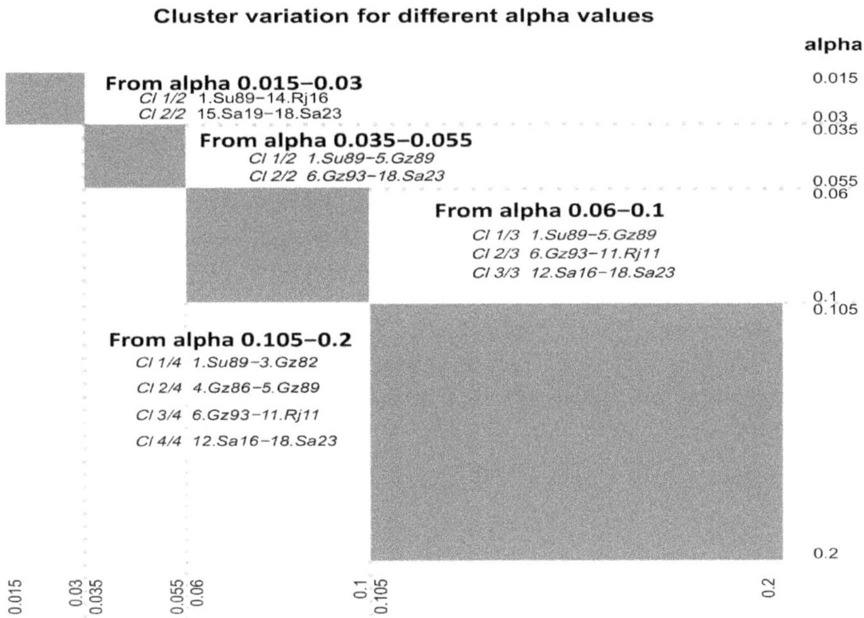

Figure 4.10: Cluster variation results for different alpha values.

As an example, a corpus containing the 18 inaugural speeches of Spanish prime ministers, delivered by six different politicians from three different parties between 1979 and 2023 is used to illustrate the application of a cluster analysis with constraints and stopping rules.

CA coordinates are used as input to cluster speeches, revealing that the first dimension is highly related to chronology when the time is included as a supplementary variable. AGNES clustering shows two groups, labeled 1979-2011 and 2016-2023, each containing three subperiods.

The first type of restriction included in the cluster analysis is the assignment of each speech to its respective political party group (must-link) and the prohibition of membership (cannot-link). This does not imply that all speeches from a particular political party are found in the corresponding political party cluster. In this chapter, the results obtained with and without constraints are compared using statistics such as the adjusted Rand Index (ARI), silhouette, cophenetic correlation and Baker's gamma correlation.

The second restriction considered is the minimum size that a cluster must have using AGNES and two variations of the divisive analysis method DIANA.

The Guttman effect is often observed in the first factorial plane of CA when one of the variables, in this case the documents, exhibits an order relationship between the modalities (the chronology of speeches). To understand the effect of chronology on cluster construction, clustering after detrended CA was initially performed.

Canonical correspondence analysis (CCA) extracts the relationships and uncovers the multidimensional structure incorporating external information known about the DTM matrix to be analyzed, in this case the chronology variable.

Once the high correlation of the cluster results with the time was verified, chronological contiguity is included as a case of contiguity-constraint analysis. A brief analysis of the reversals in clusters is presented, justifying the choice of the complete linkage method algorithm

In a cluster analysis, the number of clusters selected can exceed or fall short of the true structure. The implementation of a stopping rule in the contiguity-constraint analysis is used to automatically provide a result for this decision.

5

Textual data visualization

5.1 Introduction

Data visualization is the process of representing data or information to identify patterns, trends, or irregularities. This concept encompasses many aspects, from presentation in tabular form to the representation of multidimensional relationships that seek to reveal complex structures.

The objective of this chapter is not to provide a comprehensive review of all types of representation. Rather, it is to illustrate some representations that may be particularly useful in the analysis of textual data. The **R** packages mentioned in this chapter, with a few exceptions, will not include bibliographic references.

Data collection is a critical initial phase in any decision-making process. The tabular presentation of data constitutes a preliminary step in the analysis, allowing examination of data at a given moment, across different times, or in different places. In the context of big data, synthesizing information to identify general laws of behavior is essential. As is standard practice, the final phase of the project involved the representation of the gathered information in graphical form.

People typically make decisions in a specific order. Their priorities are as follows: graphs, followed by summary information and tables. The phenomenon of thinking through visual processing has been designated as visual thinking, which refers to the process of forming thoughts by visually processing information.

The idea of data visualization as a visual language for synthesizing information and communicating it is not new (see Marriott (2024); Friendly and Wainer (2021) for a historical review).

We are currently in the communication phase, where graphical displays are fully integrated as a primary element for conveying insights.

R has a variety of plotting functions integrated into the core system. It can also construct sophisticated plotting systems using **ggplot2** and **lattice** packages, enabling users to effortlessly create and modify existing graphics.

The document obtained can also enhance the information provided by using interactive charts (Sievert, 2020). One of the most popular libraries used to achieve this goal is **plotly**. The **ggiraph** graphic package is compatible with the **ggplot2** package, enabling seamless integration and facilitating advanced data

DOI: 10.1201/9781003397472-5

visualization capabilities. Other generic graphics are included in **highcharter** package. In addition to the functionality of zooming on a specific area, animation is also available in various **R** packages. These include the aforementioned **plotly**, **animate**, **animation**, and **gganimate**, which is used with **ggplot**.

The R Markdown document format (Xie et al., 2018) was included in the **knitr** package, which allows users to produce dynamic documents that respond to changes in the source data (Xie, 2015). The static or dynamic outputs can be a document (MS Word, HTML, LaTeX, PDF), book, slide (Powerpoint, Beamer, HTML5) or a website in different formats.

A more recent alternative to R Markdown is Quarto, an open-source system that can render documents written in different languages, such as R, Python, Julia, or Observable JavaScript. It also uses the **knitr** package and has an R package called **quarto** to set up an interface from RStudio to Quarto.

R offers a wide variety of graphics and chart creation packages. Some of these will be included in the different sections of this chapter.

Dashboards are web applications that present a set of graphics and texts in an organized manner. They can be static, where the graphics always present the same data, or dynamic, allowing users to select the data displayed through menus. They can also be installed on a personal computer.

Shiny is possibly the most popular dashboard choice among **R** users, with a wide range of applications. The software generates the necessary files (such as **R** scripts) which define the application's structure and logic, and the application can be shared with other people. Alternatively, R Markdown can be used to develop Shiny components. The company offers free restricted hosting of the generated files for web display. However, it should be noted that software such as Docker allows applications to be hosted on other servers.

The **patchwork** package has been developed for the purpose of facilitating the seamless arrangement of multiple **ggplot2** plots, which can then be utilized in a dashboard. Using R Markdown, the **flexdashboard** package allows for the creation of interactive dashboards, both independently and as a complement to Shiny. Quarto is a publishing system that facilitates the creation of dashboards for data visualization. It is gaining popularity, partly due to its ability to integrate with other software programs, including Shiny.

These representations can be classified based on the nature of the variables used, the type of relationships to be discovered, the objective, or they can be incorporated as another element within the information output provided by a statistical analysis.

This latter approach, which assigns a specific graph to each analysis, is exemplified by Blasius and Greenacre (2014) in his book on data visualization and by Blasius and Greenacre (1998) regarding categorical data analysis. As is frequently observed in research, each graph addresses a specific problem.

Another approach is to specify each type of graph according to the types and number of variables to be represented. The appropriate graph is selected based on these constraints and the intended objective (Friendly et al., 2015).

Textual data analysis also has specific requirements in this visualization process. Since the early days when computers were not available (Bertin, 2010), there has been a significant increase in processing, as well as the development of new methodologies.

The ability to access and analyze a large amount of data from the Internet, including different types of information, has led to the development of big data analysis techniques, Natural Language Processing (NLP), statistical methods, and visual tools to extract the content, summarize results, and extract patterns without losing information. If the input information is a DTM matrix with a high number of documents and vocabulary, it can easily produce an overload of information.

The comprehension and revelation of text content, patterns, or even the discovery of the unexpected requires specific representation techniques. Kucher and Kerren (2015) conducted a thorough review of visualization techniques from various publications, identifying 400 techniques that are available for consultation. These techniques can be filtered according to different criteria on the website: `https://textvis.lnu.se/`.

Cao and Cui (2016) and Hofmann and Chisholm (2020) books cover the main types of graphs applied to textual data analysis from different perspectives.

To illustrate this chapter, we will use the *Health* database described in Chapter 1 and the corpus *SpanishDisc18* in Chapter 4.

It starts with a review of visual representations in the form of tables, including charts. Chart frequencies are unidimensional representations, such as the bar plots, word clouds and distribution of words along documents. The chapter includes several two-dimensional plots that have been selected for their applicability in the use of texts, concluding with three-dimensional and animated representations.

5.2 Tabulation

Unorganized data is used as a starting point and must be structured according to the desired objectives. In many cases, the analysis of textual data is performed in conjunction with data cleaning procedures, a morphosyntactic analyzer, and word selection based on some criteria, such as frequency threshold and stopwords.

Once the database has been imported into **R**, it may not be straightforward to display the texts on screen, particularly if there are a significant number of variables. **R** packages offer a wide range of features to facilitate data management, including the ability to read external files, rename rows and columns, drag and drop, add rows and columns as needed, filter cases, synchronize data, export data subsets, and edit a data frame. Among them, we can highlight **DataEditR**, **editData**, **DT**, and **GWalkR**.

The initial phase in the process of presenting data visually is typically the tabulation, which is the process of organizing data in a tabular format.

When variables are quantitative, they are usually represented grouped together in the form of intervals (script in **chap5.a**). Interval building is also a useful technique for making frequency histogram graphs. **R** provides a basic but flexible function to obtain intervals named cut. Among the different formulas for determining the number of classes we highlight the following within the **grDevices** library, where x is the quantitative variable, N the number of cases and k the number of bins:

- grDevices::Sturges(x) considers a binomial distribution to approximate an underlying normal distribution:

$$k = log_2 N + 1.$$

- grDevices::scott(x). For normal distributions, being h the optimal bin width and S the standard deviation:

$$h = 3.5 \times \frac{S}{\sqrt[3]{N}}; k = \lceil \frac{\text{range}(x)}{h} \rceil.$$

- grDevices::nclass.Freedman-Diaconis(x). It is more robust in the presence of outliers and skewed distributions than Scott's rule or Sturges' formula. It achieves this by using the Interquartile Range (IQR) instead of the standard deviation:

$$h = 2 \times \frac{IQR}{\sqrt[3]{N}}; k = \lceil \frac{\text{range}(x)}{h} \rceil.$$

It is important to note that when N is low, the above formulas may offer very different results, especially with highly asymmetric variables. Library **abn** also includes statistics as Cencov, Rice, and Terrell-Scott but they have the drawback of overestimating k.

According to Sturges and Scott's proposal, 10 age classes are delineated using the *age* variable from the *Health* database. Freedman-Diaconis, however, suggest 11 age classes.

Figure 5.1 uses percentiles to construct the intervals (function pretty). This method imposes the constraint that the number of individuals in each class remains approximately constant, which causes the amplitudes (widths) of these classes to vary.

There are several **R** packages that allow for the creation of customized tables, such as **kable**, **kableExtra**, **formattable**, **reactable**, and **plotly**, with outputs in various formats. Other packages are **condformat**, **flextable**, **gt**, **huxtable**, **pandoc**, **pixiedust**, **rhandsontable**, **stargazer**, **tables**, **tangram** and **xtable**.

Some packages provide basic statistics in table format: (**DescTools**, **dfSummary**, **modelsummary**, **psych**, **skimr** or **summarytools**).

Sturges			Pretty			Percentiles	
Age1	Man	Woman	Age2	Man	Woman	Age3	Total
(13.9,21.4]	43	59	(10,20]	40	52	(14,18]	61
(21.4,28.8]	31	36	(20,30]	43	59	(18,22]	54
(28.8,36.2]	36	63	(30,40]	39	66	(22,27]	45
(36.2,43.6]	19	36	(40,50]	22	44	(27,31.4]	44
(43.6,51]	19	29	(50,60]	35	39	(31.4,35]	52
(51,58.4]	18	30	(60,70]	15	31	(35,42]	54
(58.4,65.8]	22	28	(70,80]	10	13	(42,48.7]	47
(65.8,73.2]	9	15	(80,90]	1	3	(48.7,57]	56
(73.2,80.6]	7	8				(57,65]	55
(80.6,88.1]	1	3				(65,88]	43

Figure 5.1: Image of an enriched table.

5.3 Chart frequencies

Representing word frequencies and document lengths is an initial step in understanding their one-dimensional characteristics.

5.3.1 Bar plot

One of the basic graphs in the study of texts is the bar plot, in which the length of the bars represents the frequency of the words. As discussed in Chapter 1 (Figure 1.2), stopwords are the most frequently occurring words in this type of representation, so it is advisable to remove them. This selection of words can also be made based on TF-IDF, keywords (Figure 1.8), characteristic words (Figure 1.9) or grammatical categories (Figure 1.5 shows only verbs). Two similar charts are the dot plot and the lollipop (Cleveland plot). The **plotly** package is an efficient tool that facilitates the creation of interactive bar plots from a graph generated with **ggplot2**. Package **ggstats** provides several tools to build diverging bar plots.

5.3.2 Word cloud

The word cloud chart is perhaps the most widespread one-dimensional representation of words in recent years. Wordle (`https://textvis.lnu.se/`) is an example of a website that allows users to generate word clouds online by entering the original text. In a word cloud, the font size is directly related to the frequency. We face some interpretation problems, such as the fact that the relationship between the words represented cannot be interpreted in terms of distance. In fact, the input matrix only contains two columns: the word and its frequency of occurrence in the text, in any order.

Another challenge in interpretation arises from overestimating the frequency of longer words compared to shorter ones, as well as the use of different colors for words. For this reason, word clouds are not generally considered a reliable tool for summarizing texts. One potential application is the comparison of one text with another. Figure 1.9, presented as a bar plot that illustrates the frequency of characteristic words used by men and women in response to question *What does health mean to you?* (script **chap5.b**). As illustrated in Figure 5.2, the word cloud displays the characteristic terms associated with men on the left and women on the right.

Figure 5.2: Word cloud for characteristic words for men on the left and women on the right.

The set.seed function in **R** can be applied to ensure the reproducibility of the same configuration of words, checking that the dimensions of the graph remain constant. However, when the graph size is modified, the algorithm recalculates the ideal positioning for each word, considering the updated dimensions of the graph.

Currently, the **wordcloud2** package supports the generation of interactive word clouds, in which hovering the mouse over a word indicates its corresponding frequency, using an HTML widget. The package **ggwordcloud** allows an easy integration into a **ggplot2** graph using the function geom_text_wordcloud as just another instruction. Textual package **quanteda** also has a function named textplot_wordcloud that reads tokenized texts in quanteda format.

5.3.3 Lexical dispersion plot

Sometimes, it can be useful to know the positions of words within a large text. The total frequency of a word can be distributed evenly throughout the text or concentrated in only one part of it. In the latter case, the repetition or concentration of words usually indicates the treatment of a specific topic. This graph is also known as an *x-ray plot* (textplot_xray function of the **quanteda** package).

As illustrated in Figure 5.3, the distribution of selected words related to equality in Spanish speeches is presented: *equality*, *discrimination* and *wom** to include *woman* and *women*. Only nine of the 18 investiture speeches discussed in Chapter 1 have been included in the plot. The script **chap5.c** contains a custom function in **R** that has been used.

It is important to note that the input data in this graph cannot be the DTM matrix, as the order of words has been lost. To determine the position of each word, we begin by tokenizing the text with the **quanteda** package.

The first three speeches, corresponding to the period we refer to as *Transition* exhibit a low frequency of these words (7, 3, and 11), with the term *equality* predominantly aligning with concepts from diverse fields. In 2004, Zapatero (9.Zp04) increased the frequency until 21 matches, focusing on the last part of his speech words related to inequality between men and women. Speech 10.Zp08 has 33 matches in a more advanced part than in the previous speech, with the majority corresponding to gender discrimination. This position is analogous to the one in Feijóo's speech (17.Fj23), though most of the frequencies of equality here refer to different issues. The Rajoy speech contains only six words (11.Rj.11).

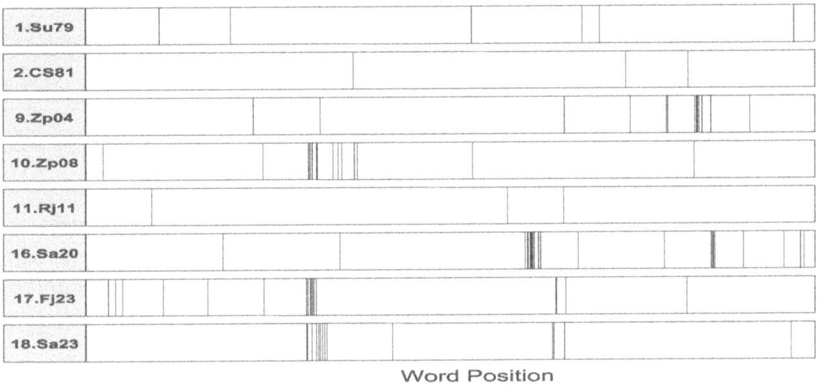

Figure 5.3: Position of targed words in documents.

5.3.4 Other charts to represent the frequency

Among the wide variety of graphs available to represent frequencies, we will highlight the following:

- Pie chart. It is a circle divided into sectors, with each sector representing a proportion of the whole based on the angles used.
- Doughnut chart is similar to a pie chart, except that it has a hole in the center. This hole can be used to include additional information or incorporate more than one variable for visual comparison. They can be obtained, for example, from PieDonut function of the **webr** package and PieChart from the **less** package, as well as with the **plotly** package for creating interactive donuts.
- Circle packing, also known as circular packing. In the unidimensional case, it is a representation similar to a word cloud, in which each word is included in a circle. Several packages include this representation, such as **ggplot2**

and **packcircles**. It can also be used to represent the same concept as a dendrogram, in which each cluster is a node containing other clusters (see geom_node_circle function of **ggraph** package).

- Waffle charts, square pie charts or gridplots represent in a small grid of squares the frequency of several categories of one variable. Package **waffle** is a utility that can be used with **ggplot2** through the geom_waffle function.

5.4 Bidimensional plots

This section includes charts that illustrate the relationship between two variables.

5.4.1 Mosaic

A mosaic chart is a type of bar chart used to visually represent a contingency table, with the objective of identifying relationships between categories. The total area is divided into rectangles to represent the frequency of each combination. The width and height of the boxes are proportional to the number of observations of each level. Conditional distributions can also be plotted using mosaics.

In practice, the data may be required in the form of a contingency table or a dataframe in which each variable is in columns, often also including the frequency of the combination of the two variables. Depending on the program, it is possible to create the graph from either type, or it may be necessary to perform transformations. The graph can be enriched by including information within each box, such as frequency and a color intensity or a gradient proportional to its size or proportional to the distance from independence using Pearson residuals.

The usefulness of textual data representation is very limited to the number of categories of the two variables that can be represented. To illustrate the mosaic plot, we will use the *Health* database to compare the relationship of a subset of words with the gender variable (script **chap5.d**). The frequencies can be observed in Figure 5.4, and the conditioned frequencies by gender in Table 5.1.

Table 5.1: Conditional frequencies by gender.

	money	job	children	family	smoke	drink	alcohol
Men	0.115	0.180	0.049	0.066	0.246	0.180	0.164
Women	0.254	0.134	0.075	0.104	0.119	0.134	0.179
Average	0.188	0.156	0.062	0.086	0.180	0.156	0.172

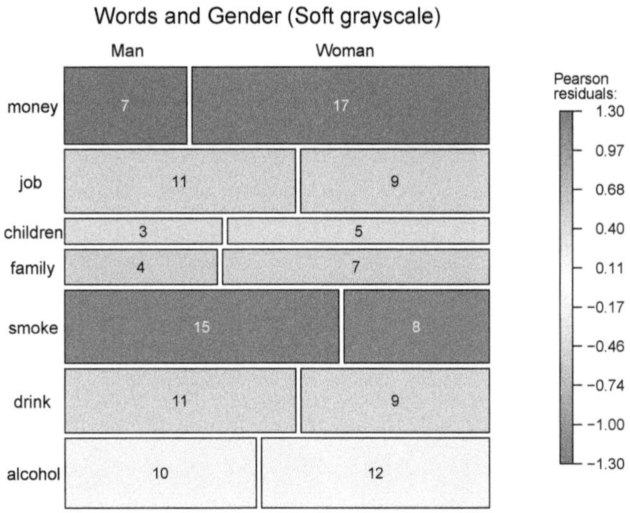

Figure 5.4: Mosaic plot of word frequency by gender.

As illustrated, among the seven words selected, *money* emerges as the most representative of women (17/24), while *smoke* is the most representative of men (15/23).

Among the libraries that create mosaics in **R,** we find the mosaicplot function in the **stats** package. The geom_mosaic and geom_mosaic_text functions of the **ggmosaic** package can be used in conjunction with ggplot charts. The **vcd** package is a solid recommendation in this case, as its name suggests, for *visualizing categorical data.*

5.4.2 Alluvial

Alluvial graphs are used to represent categorical data, with flows depicted in two dimensions to illustrate changes in structure or over time. These types of representations can be created using the **ggalluvial** package in conjunction with **ggplot**. In this data visualization, each qualitative variable is represented by a block. The height of a block indicates the size of the category, and the stream represents the frequency or size of the relationship between the categories of two blocks. The **parcats** package allows to represent interactive alluvial plots.

The following example illustrates the relationship between three words (*children*, *money*, and *smoke*) with two dichotomous variables (*Gender* and *Health status*). The example is part of the script in **chap5.e**.

The word *children* is more closely associated with women than with men, and among men, it is found more frequently among those who are unhealthy than among those who are healthy (Figure 5.5). The combination of women and *health* contains the highest frequencies of the word *children.*

The word *money* stands out among people who are not healthy and women, while *smoke* appears among healthy people and is more important among men than among women.

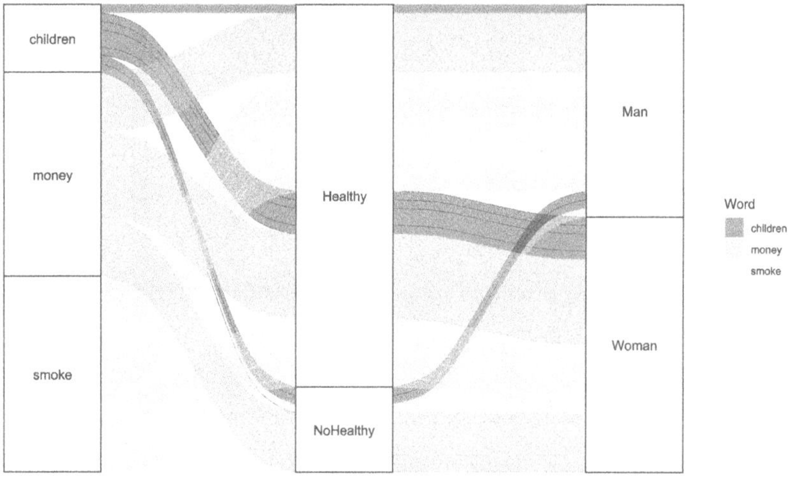

Figure 5.5: Alluvial words with *health* and *gender*.

5.4.3 Heatmap

A heatmap presents the data in a graph using colors, where the intensity of the color is related to the value that is to be represented. A heatmap plot can be presented in a grid with an equal or different number of rows and columns, or as a spatial map.

When analyzing the representation of the correlation matrix in Subsection 5.4.4, a heatmap will be included, incorporating a dendrogram from a cluster analysis. The heatmap function at the core of the **R** package allows users to specify the type of hierarchical cluster analysis they wish to perform, with extensive control over labels, font sizes, and colors. It is structurally compatible with **ggplot2**, enabling a rapid learning curve. Three additional packages are: **tidyHeatmap**, **heatmaply**, and **superheat**.

To create interactive graphics, we can use the package **iheatmapr** or those based on **ggplot2**, and then apply the **plotly** package.

5.4.4 Correlation plots

As previously mentioned, the use of techniques such as mosaics in the representation of qualitative variables is subject to certain limitations.

In 1967, Bertin (2010) applied the concept of seriation to try to solve this problem manually by moving the rows and columns of the matrix. The idea is that if there is a clear structure for organizing rows or columns, we will be able to identify blocks in the plot permuting matrices (script **chap.5e**).

When the number of categories is small, the Bertin plot displays a chart that resembles a mosaic, representing the frequencies or the distance to independence. The incorporation of numerical variables allows for the integration of a histogram within these cells.

If seriation is applied to a DTM or TDM matrix, the objective is to represent relationships between documents and words. The interpretation of these graphs becomes particularly challenging when the number of documents or words is high.

As discussed in Chapter 1, it is possible to calculate the relationships between groups of words or documents and represent them numerically. This objective can be achieved by calculating Pearson's linear correlations or cosine squares, for instance.

Current serialization packages (**kairos**, **seriation**) typically implement sorting algorithms based on the first factor derived from a SVD technique or as a result of a cluster analysis, presented as a heatmap. The package **seriation** indicates that its objective is *getting things in order*, providing numerous methods of seriation for both quantitative and qualitative variables. It also enables users to select the SVD method for ordination when starting from the original data.

Once the coordinates have been obtained by applying SVD, the relationships can be represented using graphs initially designed for quantitative variables, such as the correlation matrix (**corrplot** package), directly with the **seriation** package (seriation::pimage), **ggplot**, or **plotly**.

The Figure 5.6 has been developed with **gplots** package from the result of a symmetric matrix with the relationships between a small group of words measured using the cosine, adding the cluster dendrogram on the top (script **chap5.f**).

The word *tobacco* is closely associated with *bad* and connotations as are the terms *exercise* and *eat*. They are far away from *drink*, *alcohol*, *smoke*, and *sport*, indicating that they are perceived as separate problems. Another cluster is formed by the words *physical*, *health* and *body*.

It is also possible to observe three clusters, with each cluster consisting of two words: *job* and *healthy*, *work* and *money*, and *family* with *children*.

5.4.5 Scatter plot

A scatter plot is a graphical representation of data points for two numeric variables. The values are indicated by dots or other symbols. If the variables correspond to the coordinates of a factorial method, the two-dimensional

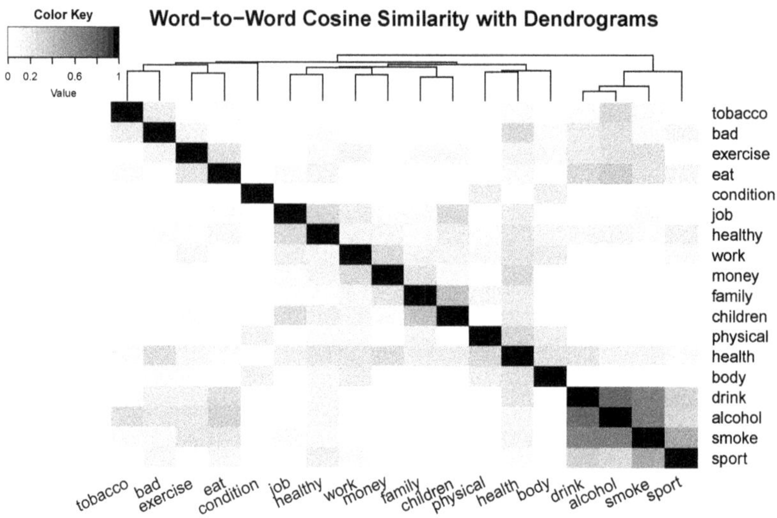

Figure 5.6: Seriation of selected words with dendrogram.

representation is usually referred to as a factorial plane. In the case of textual data analysis, the points may be accompanied by labels or omitted. Words often appear overlapped due to the short distance between the dots.

Xplortext incorporates this ggrepel possibility, exporting the plots to ggplot format to enhance plot readability (see the website address of **Xplortext** for examples of this type of graphs). The plot.LexCA function allows for the representation of the active words and documents that were taken into account when performing the CA, as well as those that were considered supplementary.

Since the number of documents or words represented as points is usually very high, it is common to set up a selection filter based on known names, position indexes, or rules based on coordinates (coord), square cosines (cos2), the absolute contribution, or meta words (words with a contribution over a number of selected times of the average word). As an alternative option, we can simply choose a certain number of the most extreme coordinates on each axis.

These representations can be enriched by incorporating contextual data, both quantitative and qualitative variables, that allow for a deeper interpretation of the axes and reveal underlying structure of the data.

If a cluster analysis is performed based on the factorial coordinates of the CA, then the barycenter of the clusters can be positioned on these factorial planes.

Xplortext facilitates this process by enabling users to perform operations based on a k-means or a *hierarchical clustering* (HCA) from LexHCca function, as well as chronological constrained hierarchical clustering from LexCHCca objects.

In some cases, the process involves illustrating trajectory arrows in alignment with an order of clusters (see Figure 4.1 for an example).

Xplortext and **FactoMineR** share similarities in their data output structure, allowing for seamless integration with **ggplot2** and **factoextra**, enhancing adaptability across diverse examples (see **Xplortext** web).

5.4.6 Bubble chart

A bubble chart is an extension of a scatter plot that allows a third variable to be included in the graphical representation by adjusting the diameter of each point. This variable can be the frequency (see Figure 5.8), and in the case of factorial methods, it can represent the absolute or relative contribution of each point.

The function geom_point of the **ggplot** package allows for the easy modification of the dot size (function scale_size) based on a variable, as well as the definition of different colors. There is a high risk of points overlapping when representing many dots. This issue frequently arises when working with texts. In this case, the level of transparency of the colors can be controlled using the alpha parameter of the geom_point function, which ranges from 0 (transparent dots) to 1 (opaque).

Interactive plots can be created by adding the **plotly** package to the above ggplot or by using the **highcharter** package. One variation is the smoothScatter function, which smoothed the dots according to frequency, allowing darker areas to be perceived as those with higher frequency.

5.4.7 Non-linear dimension reduction techniques

The utilization of dimension reduction algorithms constitutes a pivotal element in the visualization of data, as well as in the preprocessing of data for the subsequent application of other techniques.

PCA and CA are linear methods that aim to extract the structure by calculating the distance between two points in the original data and maintaining this distance in a reduced space. The orthogonality of the axes is of the utmost importance because it ensures that the dimensions are not related to each other. In other words, the position of a point on one axis is not correlated with its position on another axis.

Nonlinear dimension reduction techniques seek to favor local distances over global distances by ordering points in neighbor graphs. The two most widely used techniques for visualizing data in 2D and 3D within this approach are t-SNE (t-Distributed Stochastic Neighbor Embedding) and UMAP (Uniform Manifold Approximation and Projection). If the variables are measured on different scales, it will be necessary to scale them before obtaining a Euclidean distance matrix. The output coordinates obtained can be used to perform a cluster analysis using techniques such as DBSCAN.

A notable drawback of these techniques is that they use randomization of data, which can produce very different results. Even when a single seed is used to ensure reproducibility, adjusting the necessary parameters can introduce additional instability to the results. However, fixing a seed is often incompatible with parallel programming.

As van der Maaten and Hinton (2008) present, t-SNE's primary objective is to minimize the tendency to cluster points in the center of the map, with very weak characterization. The results are benchmarked against seven other non-linear methods. They use the Student t-distribution instead of the Gaussian distribution, but the non-parametric approach makes it hard to understand how the reduction of dimensionality is carried out, and the interpretation of the distance between clusters generally does not make much sense. Additionally, there is no guarantee that the cost function will be optimized.

The package **Rtsne** allows one to obtain 2D and 3D plots from the original data or from PCA or CA coordinates. The main hyperparameter is *perplexity*, which sets a balance between focusing on local or global aspects (see script **chap5.g**).

As illustrated in Figure 5.7, a high value of perplexity (50) yields a word representation that closely resembles a line (top-left). Perplexity of 25 and 10 show the words in the shape of a cloud. Low perplexity (3) builds small clusters of points very close, favoring a local structure. The script **chap5.g** also provides a 3D representation.

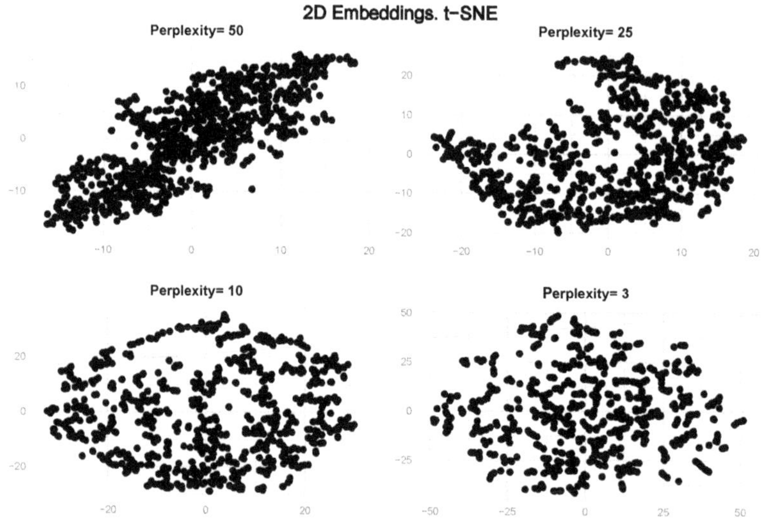

Figure 5.7: t-SNE clouds of words with different levels of perplexity.

The same analysis was performed for the *Health* data frame, selecting 145 words from the CA coordinates, using the t-SNE result as input data for the DBSCAN clustering algorithm (**dbscan** package). In this case, different values of epsilon (eps) and minimum points to build a cluster (minPts) provide groups with quite different numbers of clusters. Even when these two factors are kept constant, this big difference occurs if the seeds are different. Using dbscan::dbscan(data.dbscan, eps = 1, minPts = 3), the clustering contains 12 clusters and 107 noise points (a data point that does not belong to any cluster) (Table 5.2).

Table 5.2: Words within clusters. DBSCAN and t-SNE. Frequency in parentheses.

Number	Size	Words within clusters
0	107	
1	3	able(26), feel (126), good(145)
2	3	alcohol(22), drink(20), smoke(23)
3	4	best(31), important(58), main(28), thing(111)
4	4	care(38), oneself(22), take(37), takes(8)
5	3	(of) course(13), life(55), part(6)
6	3	cheerful(15), eager(7), work(47)
7	3	essential(11), something(17), world(8)
8	3	find(7), fit(20), general(14)
9	3	day(11), even(11), get(10)
10	3	enjoy(6), home(5), works(6)
11	3	feels(9), living(21), sport(17)
12	3	illness(59), kind(16), suffer(15)

Figure 5.8 summarizes the results of the frequency of each word in a cluster from Table 5.2 and also compares the proximity of a noise point to a cluster. The size of a bubble indicates the frequency with which a word belongs to a cluster. A small size is used for the 107 unclassified words.

An evaluation of each point will determine if modifying eps and minPts parameters would yield more substantial results. For instance, *alcohol*, *drink*, and *smoke* in cluster 2 have no close words (Figure 5.8, on the left). In contrast, the terms *illness*, *kind* and *suffer* from cluster 12 are closely linked to related words such as *disease* and *problems*. In the same way, cluster 4 (*care*, *oneself*, *take* and *takes*) is close to *person*, *mental*, and *physical*.

As t-SNE, UMAP (McInnes et al., 2018) is also used to visualize high-dimensional data in a 2D or 3D. It converts metric spaces to fuzzy topological representations, belonging to the class of k-neighbor weighted based graphs. First, a k-neighbor graph is created, assigning weights to each edge. Subsequently, a low-dimensional layout of this graph is computed. The **R** package **umap** has some parameters, as the metric (Euclidean by default, and cosine to measure the angle between two vectors when the direction is more important

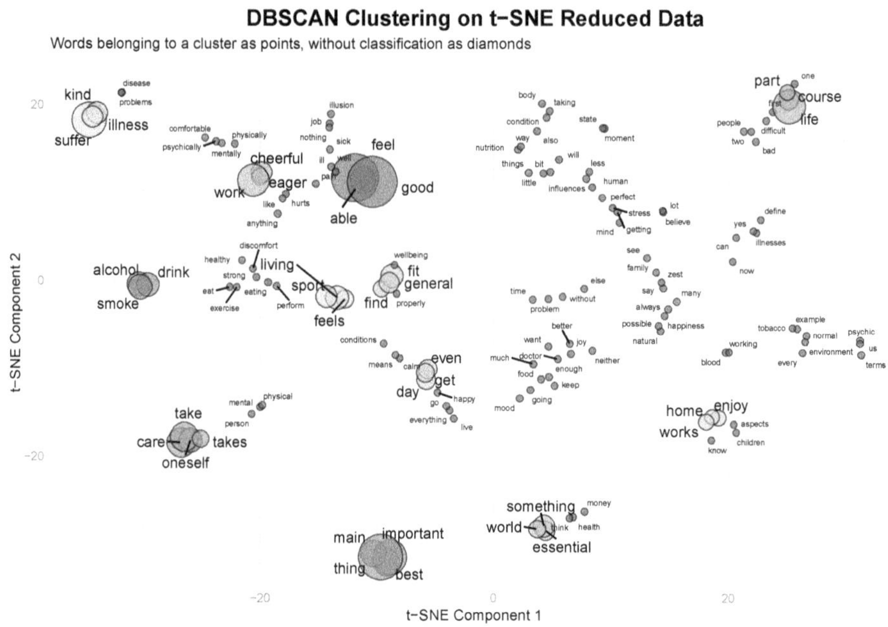

Figure 5.8: DBSCAN clustering on t-SNE reduced data.

than the magnitude), random_state to define a seed for reproducibility, and two important hyperparameters: n_neighbors and min_dist that must be carefully defined. The n_neighbors parameter determines the number of edges that come out from each node to connect with its closest neighbors. The neighbors of a given point are those that lie within a distance defined by min_dist. If this distance is low, the clusters will be compact and homogeneous.

When n_neighbors is set too low, UMAP may display multiple disconnected clusters consisting of a few small points, as well as clusters with a single point in the form of an outlier. In this case, a local structure of points is obtained. When the value is too large, the representation can be too dense, with distinct groups that are internally heterogeneous and mix with each other. In this scenario, a global point structure is extracted. From this perspective of balancing local and global objectives, the n_neighbors values that must be evaluated are those from which the randomly shaped point cloud begins to display point clusters (Figure 5.9).

As with DBSCAN and t-SNE, the results with UMAP can be improved based on the researcher's knowledge of the data structure and by testing different values in grid form to observe the changes.

Figure 5.9: Reduced map with UMAP (cosine metric, n_neighbors = 6).

5.4.8 Ellipses of confidence

The results obtained at a given point can be interpreted in terms of variability, especially with factorial methods. This can be considered an internal validation process to measure the variability. The bootstrap technique constructs contingency tables using sampling with replacement.

Lebart (2007) identifies three types of total bootstrap and compares the results by applying a CA to each bootstrap sample, using the results obtained from the original data (Alvarez-Esteban et al., 2004). The first type only corrects the arbitrary signs of the axes, considering that the same axis may appear arbitrarily reflected without affecting the interpretation. The second option considers the possibility of interchange between two axes, which can easily occur when the eigenvalues of the factors are very close. The third performs a Procrustes rotation of each configuration toward the initial configuration.

Confidence ellipses can be an essential tool for understanding the variability of data. These ellipses can be constructed for a specified percentage of data points by assuming a bivariate normal distribution of the data points, depending on the covariance matrix. Larger ellipses indicate greater variability. Another possibility is to use bootstrap techniques again, considering each row and column of every bootstrap table as elements that can be projected as supplementary (Alvarez-Esteban et al. (2010); Valencia-García and Alvarez-Esteban (2012)) (see Figure 2.5, for an example, using the ellipses function of **Xplortext** package, or plotellipses function of **FactoMineR**).

Additionally, confidence ellipses can be plotted to illustrate the variability of points belonging to a category of a qualitative variable or cluster, allowing for several forms of hulls and assuming several distributions. Examples can be found in the LexHCca and LexCHCca functions of the **Xplortext**, fviz_ellipses of **factoextra**, and ggscatter of **ggpubr**.

5.4.9 Dendrograms and clustering

From a broad perspective, a dendrogram is a tree-like representation. The word *dendro*, derived from Greek, signifies *tree*. A dendrogram is frequently regarded as the graphical representation of the relationship between objects, created as an output from hierarchical clustering. The hclust and as.dendrogram functions in the **stats** package enable the utilization of a comprehensive range of **R** libraries to depict dendrograms. Some examples of these software are: **ape**, **circlize**, **dendextend**, **ggplot**, **ggtree**, **phytools**, and **pvclust**. **Xportext** provides users with the ability to obtain figures such as *tree*, *phylo*, *clado*, *radial*, and *fan*.

Figure 5.10 illustrates these types of graphs, using the 18 investiture speeches from the *SpanishDisc18* corpus as an example (script in **chap5.h**).

Silhouette plots are also a useful tool for visualizing the quality of each cluster and the distance of each point to its cluster. However, although it should be noted that the usefulness of the plots depends heavily on the distance metric used and the agglomerative clustering method. The maximum silhouette value for a point is $+1$, indicating a perfect match between the point and its cluster. The minimum value is -1, indicating an inaccurate association. If a significant number of points have low or negative values, it may indicate an imbalance in the clustering configuration, potentially resulting in an excess or shortage of the number of clusters. Other metrics can be Davies–Bouldin and Calinski–Harabasz.

Figure 5.11 presents an average silhouette of 0.37 using the Euclidean distance, with five CA dimensions as input, the AGNES method and four clusters. Speeches 4.Gz86 and 5.Gz89 form a cohesive cluster (average silhouette 0.75) that exhibits high intracluster quality and is adequately separated from the other speeches. The most unfavorable grouping is formed by speeches 13. Rj16B, 14.Rj16C, and 17. Fj23 from the PP and the final one from Sánchez, 18.Sa23.

In certain situations, it may be useful to compare two dendrograms. This could be the case when two clustering methods have been used or when the number of input factors for performing the cluster analysis has varied.

In Chapters 3 and 4, various numerical methods for making this comparison have been outlined.

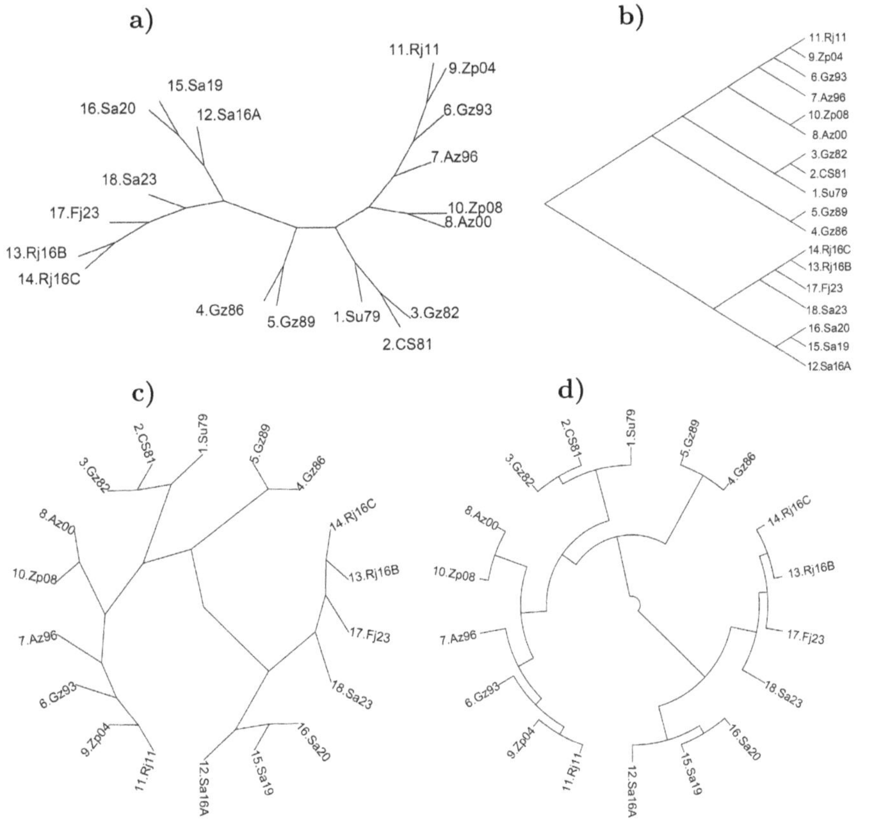

Figure 5.10: Dendrograms: a) Phylogenetic unrooted tree b) Cladogram c) Radial d) Fan.

Figure 5.12 compares two dendrograms obtained with different numbers of dimensions as input data (five and six). The function dendextend::cor.dend provides cophenetic (0.9976), Baker (0.9960), and common_nodes (0.8857) coefficients analyzing the global structure, but a visual comparison of two dendrograms when they are plotted side-by-side often reveals a greater apparent difference than there actually is.

Entanglement is a numeric measure with a value of 0 when there is a perfect alignment (no crossing lines between two dendrograms) and the maximum crossing has a value of 1. When the dendrograms are compared without alignments, the obtained entanglement result was 0.1022.

The function dendextend::untangle has seven algorithms to align the two dendrograms. The best alignment has been identified for both *step2side* and *stepBothSides* with a value of 0.0103. These are followed by *step1side* (0.0206), *labels* (0.0252), *ladderize* (0.0503), and *random* (0.0521). The worst

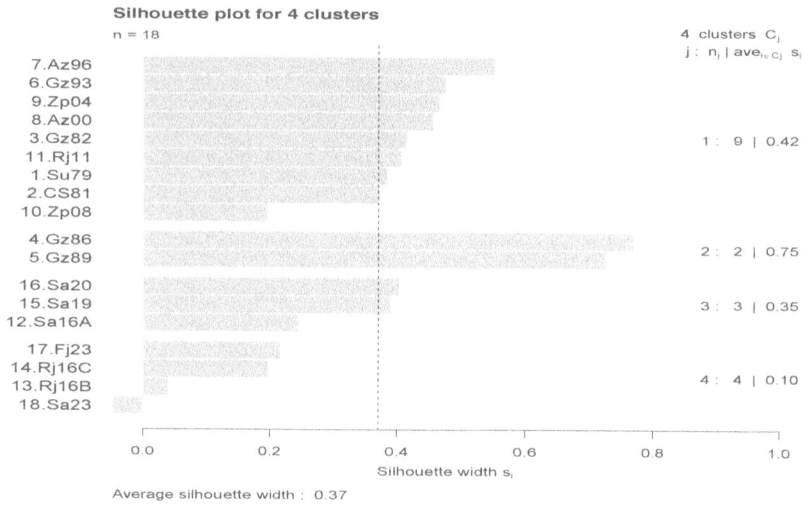

Figure 5.11: Silhouette plot for speeches with 5 dimensions and 4 clusters.

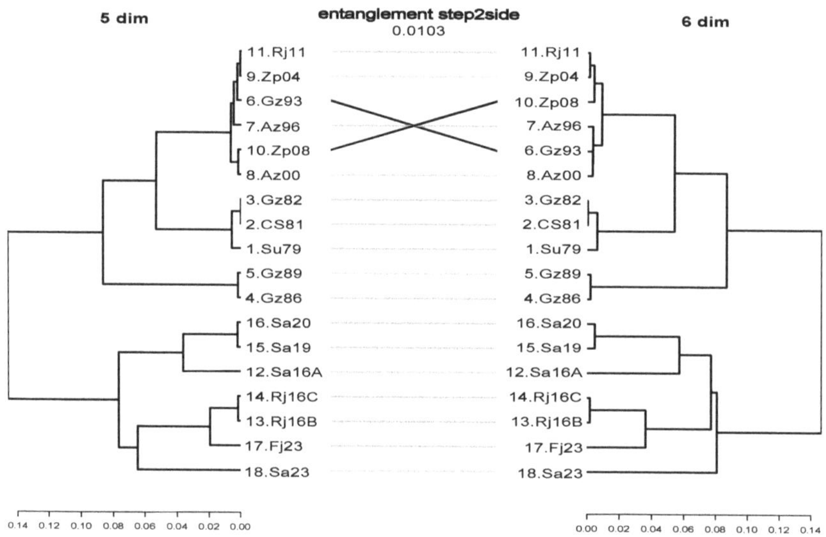

Figure 5.12: Tanglegraman for adjusted dendrograms with step2side method.

was *DendSer* (0.9755). Additionally, packages **phytools** and **factoextra** allow for the representation of tanglegrams.

The graphical representations used in Chapter 4 to relate a dendrogram to the chronological variable can be complemented graphically using the plotTreeTime command from the **ape** package. In order to facilitate comprehension of this comparison, it is necessary to initially rotate the dendrogram to be analyzed using the dendextend::rotate function, in the direction of the labels corresponding to the chronological variable. As illustrated in Figure 5.13, the dendrogram and chronology exhibit a strong correlation, with minimal intersections between the lines.

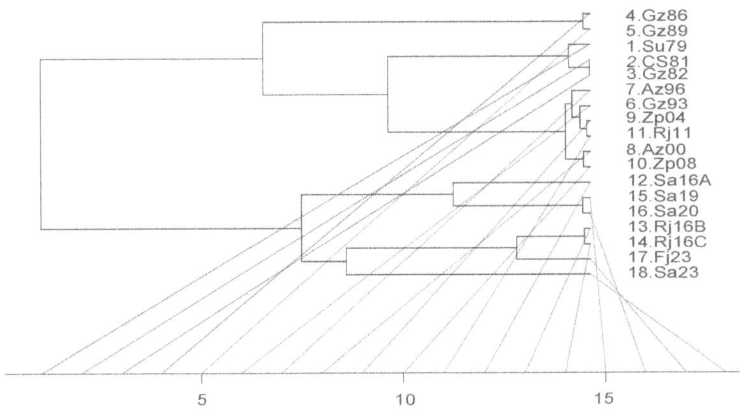

Figure 5.13: TreeTime plot with chronological variable.

5.4.10 Word tree and networks

As stated in Chapter 1, googleVis::gvisWordTree is an interactive interface between **R** and the Google Visualization API to detect sequences of words and show a hierarchical representation of the data. The objective is to place a word in context. Additionally, several network-type representations were created.

A network diagram is a useful tool for visualizing relationships between text elements. It can be used to analyze word co-occurrences (such as n-grams) or to group documents according to their similarity. The units of analysis can be either documents or terms. Networks can be simple or complex representations depending on their components and interactions. A *vertex* (or *node*) denotes a single point, representing either a document or a term. An *edge* (or *link*) signifies the connection between two vertices. A *directed edge* indicates a one-way relationship, while an *undirected edge* implies a mutual, two-way connection. Edges can also have a *weight*, which is a numerical value used to measure the strength or distance of the relationship (e.g., how similar two points are). The number of edges connected to a vertex is known as its *degree.* A sequence of edges connecting two vertices is called a *path.*

The overall process can be divided into phases. Before a network can be built, data must often be prepared. For instance, keywords or terms can be extracted from documents using methods such as RAKE, or PMI, which are available in packages like **udpipe**.

The first phase of network construction then involves using this data—or other data, like the results of a cluster analysis or a distance matrix computed from the coordinates of a CA—to define the network structure. As mentioned, the nodes (vertices) of this network can represent a word, a document, or even a concept, topic, or cluster.

The final phase is visualization. This process involves using a layout algorithm (e.g., the Fruchterman-Reingold algorithm) to determine the physical position of the nodes, followed by the graphical rendering of those nodes and the edges connecting them. In many cases, the initialization of a seed is necessary for the layout algorithm, with the objective of ensuring the reproducibility of data. This is a critical factor in maintaining the stability of the final network visualization.

In certain cases, such as with the **udpipe** package, it is possible to complete all phases within a single process.

A primary challenge in practice is that different software packages require data in specific formats. The **igraph** package, for example, offers several ways to create a network object:

- Edge list. This is usually a data frame with two columns ("from" and "to") that represent the connection between two vertices. A third column can be added to represent weights.
- Adjacency matrix. This square matrix illustrates a representation where the rows and columns correspond to the vertices. Its content represents the connection or edges between vertices. In an unweighted graph, a value of 1 indicates the presence of an edge between the two vertices and 0 if there is no edge. In a weighted graph, the numerical value represents the weight of the connection.
- Incidence matrix. This is a rectangular matrix where vertices are listed in rows and columns represent edges.

The script **chap5.i** contains a function to transform a hclust object into an igraph object. The final results are visualized using the **ggraph** package, which extends **ggplot2** for network analysis.

In Figure 5.14, the 18 speeches (documents) are depicted in a network. The positions of the nodes are determined by the Fruchterman-Reingold layout algorithm (implemented in **igraph** and used by **ggraph**). Hulls are constructed using ggforce::geom_mark_hull, and ggrepel::geom_text_repel is used to optimize the placement of the node labels.

There are several difficulties associated with this type of chart. We highlight those related to changing labels, managing labels and circle sizes, as well as their possible removal, and finally, controlling arrows and their format.

The package **plotly** is capable of constructing interactive networks from a ggplot2 object.

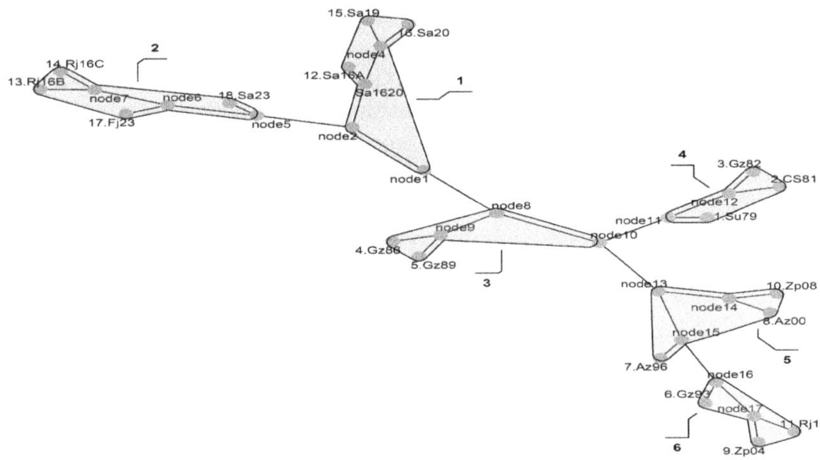

Figure 5.14: Network for 18 documents/speeches using igraph and ggplot2.

Among the packages found in **R**, we highlight the **network** package to create and modify network objects, the **networkD3** package to build interactive HTML networks using the forceNetwork function (see script **chap5.i**), the **visNetwork** package with a flexible interface that allows interactive movement of the elements, zooming, and node selection, **ggnetwork** integrating network plots into **ggplot2**, **ndtv**, and **intergraph** to convert network data between packages.

It is also important to consider the ease of exporting graphics, as well as the available formats and their quality.

5.4.11 Voronoi map and interpolation in 2D

A Geographic Information System (GIS) is a tool that displays geographically referenced data associated with specific locations, as indicated by longitude and latitude coordinates, to reveal patterns. Each point on a plane is thus defined by two coordinates and one quantitative variable. This variable can be either a contextual variable or the frequency of the point. **R** provides a variety of methods for creating maps that include spatial information, both static and interactive. The most basic plots are dots, which have a fixed size, and bubbles, whose sizes are proportional to the size, with the option of adding colors.

Choropleth maps are one of the most common thematic representations that use different shades and colors to indicate changes of values in predefined areas. It is possible to divide one plane into regions, starting at the coordinates of the points in two dimensions. Each potential point within a specified region is closer to the actual point defining that region than to any other actual point.

One method for constructing these regions is the Voronoi (or Dirichlet) tessellation (Okabe et al., 2000). A potential challenge that arises in GIS systems involves the possibility of multiple points having identical coordinates. Depending on the objectives, this problem may be addressed through the removal of duplicates, the addition of frequencies, or the calculation of the average or the sum. If necessary, a new label can be generated for each point by combining the labels of the points (document names or word names). To ensure the avoidance of duplicates, a small random value can be added to guarantee that no two points have identical coordinates (jittering procedure). In this case, it is essential that the same seed is applied to ensure the reproducibility of the analysis.

The script **chap5.j** starts extracting 811 words from the *Healt* base, removing stopwords and numbers, regardless of their frequency. From these words, 30 related to health conditions and symptoms have been selected. A total of 98 active words were identified, including those with a minimum frequency of eight. Among these active words, there are 10 that belong to the group of 30 selected words. Consequently, of the 30 words selected, 20 are regarded as supplementary when performing a CA, and 10 are considered active. The coordinates of these words are now combined into a new file, which includes both 98 active words and 20 supplementary words.

Duplicates were detected at two coordinates (where *fever* and *malfunction* shared one location, and *aches* and *swollen* shared another). A jitter procedure was applied, and all 30 words were retained.

The next step is the transformation of the selected data frame with the coordinates into an sf object of the **sf** package (Pebesma, 2016). From a practical standpoint, it is more advisable to perform this import with the duplicate points and then carry out the desired processing to remove duplicates. The import includes 116 words and is performed without a CRS (coordinate reference system). As the objective is to represent the 30 selected words, the first step is to calculate the map limits in the event that all 116 words are represented.

As illustrated in Figure 5.15, which was constructed with the jitter option enabled, the 30 selected words are predominantly located in the third quadrant. The use of colors also allows us to observe that the areas corresponding to *illness (59)*, *ill (36)*, *pain (45)*, and *problems (28)* are the most frequent. A small area indicates that the word is in a lexical profile similar to the adjacent words.

When the number of points is high, representations using Voronoi diagrams can be confusing and visual quality is greatly reduced, as the area of the cells will be small.

Kriging interpolation methods are advanced GIS techniques that enable the frequency or contextual variables to be maintained within the representation. It generates a continuous surface of values based on the distance to nearby points, while also considering the spatial autocorrelation of the data.

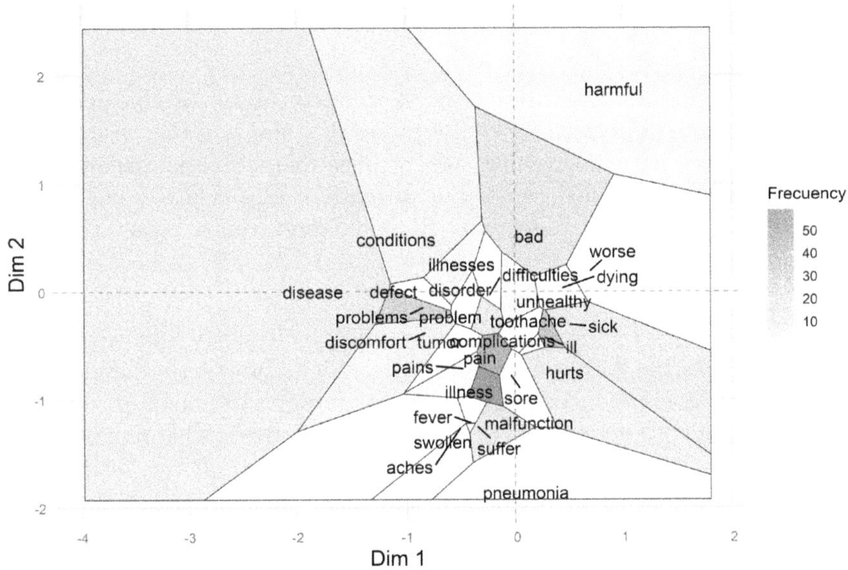

Figure 5.15: Voronoi diagram. Words related to health conditions and symptoms.

Spatial autocorrelation refers to how the values of the points relate to each other based on their distance.

Kriging is a procedure that involves two primary decisions: first, fitting a theoretical variogram model to the data, and second, selecting the appropriate kriging method.

First, to describe the spatial autocorrelation of the data, a model must be fitted to the experimental variogram. This involves establishing several key parameters:

- The type of the model: Spherical, Exponential, Gaussian or Stein's.

- Kappa. This is a smoothness parameter specific to Stein's/Matérn models that allows for a variable level of smoothness. A kappa value of 0.5 demonstrates a high degree of similarity between the results and those of the exponential model. As this value decreases, the representation becomes rougher and less smoothed. A very large kappa value exhibits a high degree of similarity to the Gaussian model. Intermediate kappa values ranging from 1.5 to 2.5 can offer valuable insights into smooth variations.

- Nugget. The error at a distance of zero, indicating noise, measurement problems, or short-range variability.

- Psill. The partial sill represents the variance of the correlation process once the nugget effect has been eliminated. Sill is the variance point at which autocorrelation ceases (Nugget + Psill).

- Range. The distance at which the curve flattens out, reaching the level of the sill.
- Log transformations. Logarithmic transformations of the variable being predicted are often performed for the purposes of stabilizing the variance (homocedasticity), normalizing the data, or linearizing the relationships. This transformation is applied *before* the variogram is calculated, and the resulting structure (key parameters) will be different from a variogram of the untransformed data.

The **automap** package includes a function called autofitVariogram that identifies the optimal parameters to minimize RSS (Residual Sum of Squares) error and obtain the corresponding variogram. The function autoKrige then uses these parameters and applies the most appropriate model to compute the maps of the prediction and the variance (or errors). This function supports several types of kriging, which are selected based on the formula and arguments provided:

- Ordinary kriging. It is a method of analysis used when the mean of the variable is constant but unknown within the specified area.
- Universal kriging. It is used when the model has a spatial trend.
- KED. If external variables can be used to define the trend, then the model is known as Kriging with an External Drift (KED).
- Block kriging. When the objective is not to predict the exact location of a point but rather to estimate an area, block kriging can be a more suitable option.
- Cokriging. It is used to perform interpolations by using the correlation between different variables together.

However, it should be noted that kriging interpolation does not always provide good results, particularly when the number of points is limited or when the points are not spatially autocorrelated.

Figure 5.16 shows the predicted frequency using ordinary kriging based on the CA coordinates of the 116 words in the Voronoi example. The colors of the forecast map were modified by adding an alpha transparency value to each prediction. This means that points with higher prediction variance are less intense than points with lower variance at the same forecasting value.

5.4.12 Other representations

The representation of textual data depends on the objectives, number of documents and words, document length, and type of contextual variables used. Without intending to be exhaustive, we will point out a few.

The script **chap5.k** includes examples of biplots, circular packing, and sunburst representations.

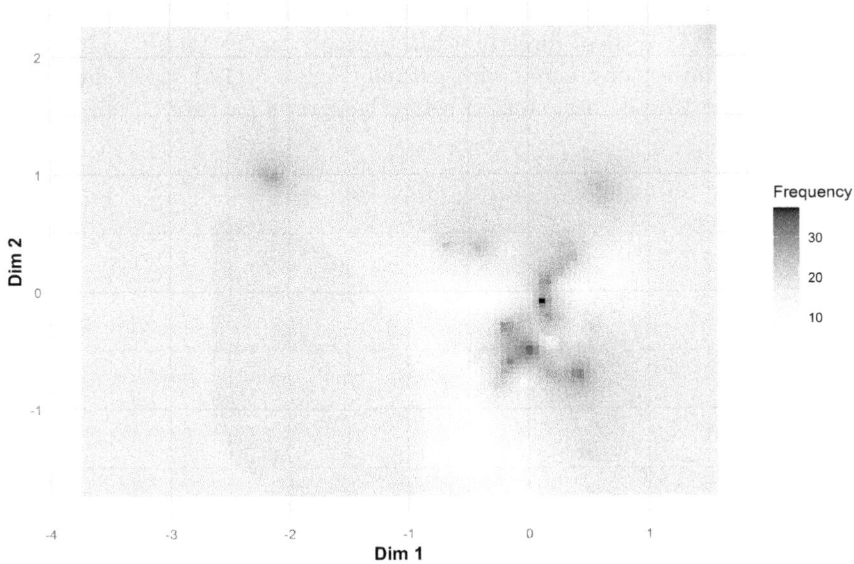

Figure 5.16: Frequency prediction on a factorial plane using ordinary kriging.

5.4.12.1 Biplots

As outlined in Chapter 2, documents and words are represented separately based on their coordinates in factorial planes. If documents and words are represented together on the same factor plane, please note that it is possible to interpret the distance between row points and between column points. However, this does not apply to the distance between row points and column points. One effective solution to this problem is to create an asymmetrical plot to present the columns in the row space or vice-versa. The standard coordinate S can be computed from the principal or original coordinate as: $S = P/\sqrt{eigenvalue}$.

The function fviz_ca_biplot of the **factoextra** package offers a variety of options for representing both symmetrical and asymmetrical plots using the coordinates of a CA.

5.4.12.2 Circular packing

Circular packing, also known as a circular treemap, provides a less rigid way to visualize a hierarchical structure than the standard treemaps (see **treemap** and **d3tree** for treemaps in **R**). There are two approaches to constructing this plot: retrieving the hierarchy from a dendrogram or defining the data from a dataframe.

Several **R** packages can create these plots. For static versions, **ggraph** is a popular choice (often using **packcircles** internally for the layout), allowing the circle size to be set proportionally to a variable. For interactive plots, **circlepackeR** creates them directly, while **ggiraph** can be used to add interactivity to a static plot created with **ggraph**. Figure 5.17.a) shows a circular packing for the 18 speeches, created using the **ggraph** package.

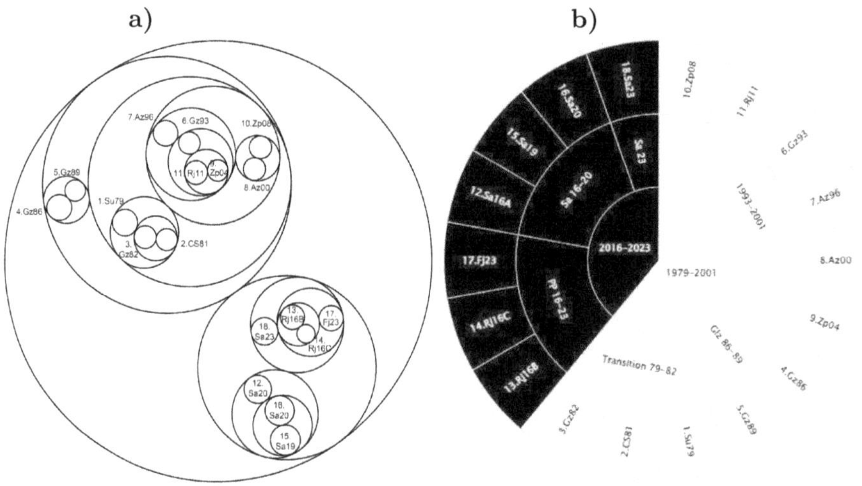

Figure 5.17: Other plots a) Circular packing b) Sunburst.

5.4.12.3 Sunburst

A sunburst chart is a type of data visualization that organizes hierarchical data into concentric rings. The top-level category is represented in the innermost circle, and each outward ring represents the next level in the hierarchy. The segments of each ring show the breakdown of the parent category.

The **highcharter** package can be used to create interactive sunburst charts, which allow for labeling each segment of the rings. As illustrated in Figure 5.17.b), a sunburst plot is presented using the speeches example once more.

5.4.12.4 Spider

The radar, or spider plot, is a two-dimensional chart used to compare several quantitative variables. It plots the values for different categories of a qualitative variable as distinct polygons on the same chart. Each quantitative variable is given its own axis (or "spoke") radiating from the center. The values for each category are plotted on these axes and then connected by lines, forming a polygon. This allows for a quick comparison of the profiles of different categories across the same set of variables. The **fmsb** package implements this plot.

5.4.12.5 Arc diagrams

While many network diagrams place nodes freely in two-dimensional space, arc diagrams visualize a network by organizing all nodes along a single linear axis. The connections (edges) between these nodes are then drawn as arcs. This layout is particularly useful when the linear ordering of the nodes is significant, such as in a timeline or a sequence.

The object can be create using the graph_from_edgelist from the package **igraph**. Subsequently, the **ggraph** can be used with the layout specified as 'linear'.

5.4.12.6 Chord

Chord diagrams are a useful tool for visualizing relationships and flows between a set of elements, showing how these components interact. The **circlize** package includes the chordDiagram function to create this plot. One common application is representing the relationships in a correlation matrix.

5.4.12.7 Self Organizing Map

A self-organizing map (SOM) or Kohonen network is a type of artificial neural network that aims to use a low-dimensional representation of high-dimensional data, allowing the points to be visualized. On a two-dimensional map, points with similar values for the variables will appear close together and are grouped. Possibly the two most widely used packages in **R** are **som** and **kohonen**.

5.5 3D and animated plots

The usefulness of 3D graphics depends on the objective. They can be used for static presentation—to synthesize and display a known data structure—or as an interactive research tool for data exploration. In the latter case, the ability to interact with the plot is crucial for the discovery process.

Most publications use static graphs, which are a type of visual representation of data that doesn't change. **R** offers a variety of packages that facilitate the representation of data in three dimensions, with the option to incorporate interaction and storage in HTML format. These packages can be easily integrated into web pages or browsers, enabling seamless data visualization. Additionally, **R** provides a comprehensive solution for generating movies, typically by creating an animated GIF file.

Among the graphics packages in **R**, we would first like to highlight **rgl**, which represents points, spheres, and text in 3D, with great control over colors and the ability to rotate and zoom in on any part of the graph. The exportation of graphics can be achieved in a static manner using the **rgl** package, with the resulting images being saved in the PNG and PDF formats. Additionally, the rglwidget function facilitates the exportation of dynamic graphs in HTML format, leveraging the **htmlwidgets** package for this purpose (see the script **chap5.1** for an example representing the Spanish speeches). It is also possible to program an animation by specifying the axes to be rotated and the duration of the animation using the rgl::play3d function. The movie3d function in the **rgl** package stores of animated graphics as GIF files, if the **magick** package is installed. Otherwise, the individual frames will be stored separately (e.g., as GIFs), and they must be joined outside the **R** environment using other software.

The **plotly** package is an excellent resource for creating interactive 3D graphs and storing the results in HTML files using the **htmlwidgets** package.

A DBSCAN cluster analysis has been performed using the same 18 points of Spanish speeches. Figure 5.18 shows an interactive 3D plot summarizing a grid. The three axes represent the input parameters: epsilon (eps), minimum points (MinPts), and the number of CA dimensions. The result of each combination is mapped to visual aesthetics; for example, the number of clusters is mapped to color intensity.

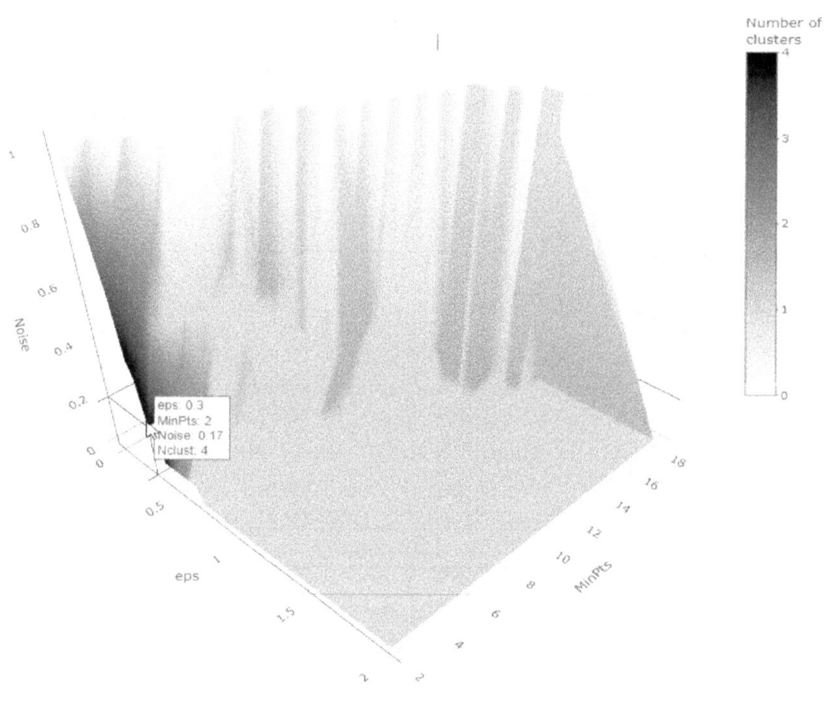

Figure 5.18: 3D plot to chose some parameters in DBSCAN.

At the top of the figure, in white color, there are no groups, corresponding to small eps values and a noise value equal to 1. These eps values can increase as minPts also increases without producing groupings as represented by the white color. As illustrated at the bottom of the figure, for large values of eps, it is evident that all individuals are grouped into a single cluster. In the graph, the combinations for which the number of clusters is higher (four) are represented in black on the left side.

The point indicated by the mouse cursor shows that for eps of 0.3 and MinPts of 2, a noise level of 0.17 (17%) was obtained, corresponding to three speeches that were not grouped. This configuration resulted in a total of four clusters (excluding the noise points).

Other packages are **ndtv** (NetworkDynamicTemporal Visualizations), **threejs** and **animation**.

5.6 Conclusion

Data visualization is no longer just the process of representing data or information with the aim of synthesizing large volumes of data. It has been incorporated into the decision-making process to identify structures, trends, or irregularities, becoming an additional phase of the investigation.

Language **R** offers a wide range of packages for the developing sophisticated graphics, encompassing static, dynamic, interactive, and animated formats. In the contemporary digital landscape, interactive tools are particularly pivotal. Software such as **Shiny** package enables the development of web-based applications and dashboards, allowing for the creation of dynamic graphics that can be integrated within various documents to enhance their impact.

This chapter has aimed to provide an overview of the main data visualization tools, starting with tabulation and one-dimensional representation, which has been enhanced by the ability to add information interactively. It maintained the traditional classification of two-dimensional plots for representations relating two variables. However, it also highlighted the increasing trend of using two-dimensional graphs to relate many variables, with one example being the use of heatmaps combined with dendrograms from cluster analysis.

We are aware of the complexity involved in creating many graphical representations, especially in the case of kriging analysis. Integrating packages within the **ggplot2** and **plotly** saves a great deal of work, allows code reuse, and facilitates the learning curve.

Finally, one aspect of crucial importance is that the software used has strong capabilities for exporting results in different formats and with high quality.

6

Textual data and contextual data playing a symmetric role

6.1 Introduction

In textual analysis, documents are often accompanied by supplementary information that clarifies and enriches their meaning. Most textual databases archive contextual and free-text data. Scientific article databases, for example, collect textual data such as titles, keywords, and abstracts, as well as contextual data in the form of continuous or categorical variables such as publication dates, journal titles, and authors' names. One such database is *Web of Science*. Similarly, legal databases archive court decisions alongside textual data such as the legal and factual basis and contextual data such as the decision date, crime or offense code, court, and judge names. Questionnaire-based surveys are another example. Addressing certain issues with a series of closed questions yields quantitative or categorical variables. Adding one or two open-ended questions yields free-text responses.

An interesting case occurs in questionnaire surveys when a topic is examined from multiple angles. This approach uses both closed and open-ended questions. The former tend to be more numerous and lead to quantitative and/or categorical variables. Open-ended questions will usually be fewer in number. The objective is to group respondents into homogeneous clusters with regard to the topic.

Therefore, we plan to proceed as follows: First, we will preprocess the data using a factorial method. Then, we will cluster the data based on the informative factorial axes. Multiple Factor Analysis for Contingency Tables (MFACT; Bécue-Bertaut and Pagès, 2004, 2008) is a suitable preprocessing method for mixtures of quantitative, categorical, and textual data. MFACT is an extension of multiple factor analysis (MFA; Pagès, 2014; Escofier and Pagès, 2016) that includes frequency data. Chapter 3 presented clustering methods appropriate for textual data and ways to exploit their complementarity with factorial methods. Here, we will use a combination of hierarchical clustering and the k-means process to consolidate the initial partition.

This chapter provides an overview of MFACT. Section 6.2 presents the example used throughout the chapter. Section 6.3 sets out the data structure and notation. Section 6.4 recalls the principles of MFA and how it is extended

DOI: 10.1201/9781003397472-6

to contingency tables in MFACT. Section 6.5 explains how clustering is used in this case. Section 6.6 briefly discusses other types of applications that benefit from the MFACT method because of their combination of numerical and free-text scoring. Section 6.7 presents and discusses the results obtained using the example. A short conclusion ends the chapter in Section 6.8.

6.2 Example: children and reading survey

To learn more about children's reading practices and attitudes, N. Rajadell conducted a study in 1990 in a town on the outskirts of Barcelona. She aimed to identify and characterize the factors that motivate children to become good readers. Fifth-grade students (ages 10-11) were surveyed using a questionnaire that included closed questions about their reading habits (see Table 6.1) and two open-ended questions that solicited more spontaneous responses (see Table 6.2). Additional information is available regarding academic achievement, particularly the verbal skills score, used below, as well as family socioeconomic characteristics, which are not used here. In this chapter, we use a simplified version of the data, retaining only three closed questions. Our goal is to demonstrate the nature of the results and explain how they are interpreted. Additional results can be found in Bécue-Bertaut and Pagès (2008). The free-responses have been translated into English.

Table 6.1: Closed questions about reading habits.

1. At home, we have	(few, enough, many books)
2. I read	(a little, a fair amount, a lot)
3. I read with	(great difficulty, some difficulty, ease)

Table 6.2: Open-ended questions about reading habits.

1. Para mí, leer es... (For me, reading is...)
2. Creo que leer es importante porque... (I think reading is important because...)

Table 6.3 illustrates the diversity of the children's responses and sheds light on what attracts them to reading. One of the least motivated children said *It gets boring because you always have to look at the letters and stuff* while one of the most enthusiastic children answered *I don't know how to explain it, but I like reading.* The second open-ended question steers the responses toward

more academic aspects. However, this did not stop one avid reader from saying that reading is important because *without reading, I couldn't travel.*

Table 6.3: Some of the responses to the two open-ended questions.

Resp.	**Q1** *"For me, reading is..."* **Q2** *"Reading is important because..."*
A101	**Q1** Fantastic! I like to read a lot and then tell what I have read. If I read every day, when it is my turn to read in class, I will know how to read quickly. **Q2** If not then at the end of the class I will not know how to read well.
S554	**Q1** It gets boring because you have to keep looking at letters and stuff. **Q2** Because you learn to see how words are spelled and everything.
R514	**Q1** It is very interesting because you can go to the world of fantasy. For example, you can travel to the Middle Ages or be in prehistory. **Q2** Without reading, I could not travel.
R404	**Q1** Fun and interesting. **Q2** I learn new things.
R181	**Q1** Like living a different kind of life, I learn, I have fun and I give myself. I don't know how to explain it but I like to read. **Q2** You learn new things.

6.3 Data structure and notation

6.3.1 Multiple data table

Table 6.4 displays the multiple table constructed from the data. A complete disjunctive table is associated with each active closed question. Each open-ended question is associated with a DTM. If this type of variable were included in the survey, a quantitative table would also be considered.

Table 6.4: Multiple table for analysis based on the example.

Resp.	Closed Questions	Open Q1 For mi to read	Open Q2 Reading important
	Indicator var.	Word counts	Word counts
I=816	K_1 categories	K_2 word col.	K_3 word col.
	K_1=9	K_2=77	K_3=77

6.3.2 Notations

J sets of variables are observed on I respondents: J_c sets of categorical questions (one set of categorical questions = one set of column indicators), J_f open-ended questions (one open-ended question = one set of column words), and J_q sets of quantitative questions (one set of quantitative questions = one set of quantitative columns). J=J_c+J_f+J_q is the total number of sets of the multiple table. Regardless of the set type, j refers to a set and k refers to a column. K_j is the number of columns of the table associated with the set j. K=$\sum_j K_j$ is the total number of columns of the multiple table.
Each set of categorical questions is associated with a respondents × indicator variables table. Each open-ended question j is associated with a DTM or respondents × words table. Each quantitative set is associated with a respondents × quantitative variables table. The multiple table, or global table, juxtaposes all these tables.

The table corresponding to the responses to the closed categorical questions shows whether or not respondent i exhibits the kth column category in set j (x_{ikj} = 1 or 0). Categorical variables can be structured according to J_c groups. In this example, all closed questions are included in the same set, so $J_c = 1$. For each of the J_f contingency tables, associated with one of the open-ended questions (here, $J_f = 2$), p_{ikj} indicates the proportion with which respondent i uses word column k to answer question j, proportion calculated between the number of occurrences pronounced by respondent i of word k in set j (=y_{ikj}) and the grand total $N = \sum_{ikj} y_{ikj}$ over all the contingency tables. Thus, we have that $\sum_{ikj} p_{ikj} = 1$. Additionally, in accordance with standard practice, the margins of the global table $\mathbf{P}$ and its subtables, noted $\mathbf{P}_j$, are defined by replacing the summation index(es) with a dot. The general term for the column margin of $\mathbf{P}$ is $p_{i..} = \sum_{kj} p_{ikj}$, and for the row margin, $p_{.kj} = \sum_i p_{ikj}$. The column margin of table j, as a subtable of the global table is $p_{i.j} = \sum_k p_{ikj}$, and the sum of all terms of each $\mathbf{P_j}$ is given by $p_{..j} = \sum_{ik} p_{ikj}$. If any quantitative variables are considered, the table corresponding to the responses to the closed quantitative questions shows the value z_{ikj} taken by respondent i for k^{th} column variable in set j. These quantitative variables could also be organized into J_q groups. In the example used in this chapter, there are no closed questions, so $J_q = 0$.

6.4 Multiple factor analysis for a mixture of quantitative, categorical, and frequency data

Our goal is to combine three types of data—categorical, quantitative, and frequency— into the same analysis. First, we provide an overview of classical MFA. Then, we demonstrate how to adapt this method to analyze multiple contingency tables with different internal column margins. The result is the MFACT method. Lastly, we will discuss how to analyze all three types of data simultaneously.

6.4.1 MFA principles

MFA (Pagès, 2014; Escofier and Pagès, 2016), is used to analyze a set of statistical units (here, respondents) described by J sets of variables, which are initially assumed to be quantitative. The process is as follows. First, a principal component analysis (PCA; Husson et al., 2009, Chap 1) is performed on each set of quantitative variables separately. Typically, both metrics are unitary, but others can be selected. The sets may be standardized according to the usual criteria. This allows the first eigenvalue of each separate analysis to be calculated. Next, a non-standardized PCA is performed on the global table. For this analysis, the columns are reweighted to balance the influence of each set on the first principal axis. This is achieved by dividing the initial weight of each column variable, which is usually uniform, by the first eigenvalue obtained from the PCA of the respective set. For categorical data, the appropriate method is multiple correspondence analysis (MCA), a factorial method designed for this type of data (see Husson et al., 2009, Chap 3). Note that when starting from a complete disjunctive table and using the correct column weights, MCA is equivalent to non-standardized PCA (Bécue-Bertaut and Pagès, 2008). Therefore, MFA applied to quantitative and categorical sets is still a type of PCA.

MFA outputs standard factorial analysis results and tools for comparing respondent structures generated by each set of columns:

- As with any factorial analysis, MFA provides a global representation of the rows and columns which is called the *global structure.*
- Specifically for the study of a multiple table, MFA provides a superimposed representation of the structures induced by each set, called *partial structures*, as well as a global representation of all the factors from the separate analyzes.

6.4.2 MFA for contingency tables

Multiple factor analysis for contingency tables (MFACT; Bécue-Bertaut and Pagès, 2004, 2008) is an appropriate method for analyzing multiple lexical tables. This method combines internal correspondence analysis (ICA; Benzécri, 1983; Escofier and Drouet, 1983; Cazes and Moreau, 2010) and MFA. ICA accounts for differences in column margins, while MFA balances the influence of the different tables in the global analysis. MFA-type results are provided. The basic principles are presented below.

In short, the lexical tables are modified so that the initial generic term p_{ikj} is transformed into:

$$\tilde{p}_{ikj} = \frac{p_{ikj} - \left(\frac{p_{i.j}}{p_{..j}}\right) \cdot p_{.kj}}{p_{i..} \cdot p_{.kj}} = \frac{1}{p_{i..}} \left[\frac{p_{ikj}}{p_{.kj}} - \frac{p_{i.j}}{p_{..j}}\right].$$

This transformation means that, instead of comparing column profiles (i.e., word profiles) with the global column profile (i.e., the mean word profile), comparisons are made with the marginal column profile of their respective tables (i.e., the internal margin column profile). The differences between the two are expressed in square brackets in the former formula. This corresponds to the ICA approach.

The influence of the subtables on the first global axis is balanced by overweighting each column of table j by the inverse of the first eigenvalue, which is calculated in a separate analysis of that subtable. Thus, the weight of column kj is equal to $\frac{p_{.kj}}{\lambda_1^j}$. To ensure consistent row weights across all analyses, the weights are derived from the global row margin, i.e., $p_{i..}$. This applies to both the separate analyses and the global analysis. This corresponds to the MFA approach.

We then perform an overall analysis consisting of a non-standardized PCA of the global table, which juxtaposes all the contingency subtables. This analysis takes into account that:

- The general term of the global table is $\tilde{p}_{ikj}$ defined above.
- The weighting system for the rows (and the metric in the column space) is $p_{i..}$; $i = 1, \ldots, I$, where $p_{i..}$ is the average relative weight of the rows across all contingency tables.

- The weighting system for the columns (and the metric in the row space) is $\frac{p_{.kj}}{\lambda_1^j}$; $k = 1, \ldots, K_j; j = 1, \ldots, J_c$.

As with MFA, graphs and interpretation aids are provided.

6.4.3 Mixture of active subtables of categorical, frequency, or quantitative types

Now, we will assume that we are dealing with a multiple table consisting of active subtables of the categorical, frequency, or quantitative type. In order to include all three types of data in the same analysis, we will treat the indicator variable tables derived from the categorical variables as we would in MCA and the contingency tables derived from the frequency columns as we would in ICA. Quantitative tables will be treated as they would be in PCA. In each analysis, the rows will have the same weight, equal to $p_{i..}$.

Next, the MFA approach is used, which is equivalent to performing an non-standardized PCA on the table that juxtaposes all the subtables (Bécue-Bertaut and Pagès, 2008) with:

- The weight of the respondent-rows (and the metric in the column space) equal to $p_{i..}; i = 1, \ldots, I$, where $p_{i..}$ is the average relative weight of the rows across all frequency tables;
- The weight of the columns in set j (and the metric in the respondent space) equal to the initial weight of the column divided by λ_1^j.

This global analysis yields the following results:

- a global representation of rows (respondents) and columns (quantitative variables, categories and words).
- a superimposed representation of the structures induced by each open-ended question and each group of closed-ended questions, called *partial structures*, as well as a representation of the factors of the separate analyzes.

The MFA's many interpretation aids make understanding the results easier. These aids are presented in Section 6.7.

6.5 Completing MFACT with a clustering step

Respondent clustering can be performed using their coordinates. Only the informative axes are retained. This filters out the noise transmitted by the remaining axes. The clustering step itself offers no new methodology because the problem is the same as with any clustering based on quantitative data.

However, we emphasize the importance of combining MFA with clustering because this process efficiently clusters data consisting of a mixture of categorical, frequency, and quantitative data. This issue arises in several applications, particularly in textual analysis.

Hierarchical agglomerative clustering typically uses the classical Euclidean distance between respondents and Ward's criterion to successively group intermediate nodes. Ward's criterion is particularly appropriate when clustering complements a factorial method. After cutting the hierarchical tree to obtain a k-cluster partition, it is advisable to consolidate the partition to maximize the inertia ratio:

$$inertia\ ratio = \frac{between\text{-}cluster\ inertia}{total\ inertia}.$$

This is achieved through a direct partitioning algorithm such as k-means, which iteratively reassigns each respondent to the cluster whose CoG is closest to them. The result is clusters that are more homogeneous and better separated.

6.5.1 Partition characterization

The selected partition can be fully understood through its statistical description, which considers all available information about the respondents, regardless of whether that information played an active role.

First, the clusters are positioned on the factorial plane using either their CoG alone or by displaying all respondents differentiated by color according to their cluster. This spatial representation shows the clusters' relative positions and distances from each other, as well as how the clusters differ in terms of the factorial results. Their proximities and/or oppositions are interpreted according to the MFA results.

Then, the association between the partition and both the closed variables and the open-ended questions is measured as follows:

- For the quantitative variables, that were either active or supplementary in the previous factorial analysis, the square correlation ratio η^2 is used.

- For the categorical variables, the Cramer's coefficient V, a standardized form of the χ^2 statistic is used. This statistic is calculated using the table that crosses the partition and the categorical variable as:

  $$V = \left(\frac{\chi^2}{N \cdot \min\{(I-1), (J-1)\}} \right)^{\frac{1}{2}}.$$

  This statistic allows for comparisons between tables with different totals and dimensions. It varies between 0 (complete independence between rows and columns) and 1 (maximum association between rows and columns).

- As for the open-ended questions, the Cramer's coefficient is calculated using the DTM that crosses the partition and vocabulary.

6.5.2 Cluster characterization

The clusters are characterized by the respondents' features and lexical specificities. The first are identified by the responses to the closed-ended questions either active or supplementary. The clusters' lexical features are captured through characteristic words and open-ended responses.

Thus, we extract and list the following:

- Quantitative variables with a significantly higher or lower mean in a given cluster than the overall mean.
- Categories whose proportions in a given cluster differ significantly from the overall sample proportions.
- Words that are significantly overrepresented or underrepresented in a given cluster compared to the sample as a whole. These words are called characteristic words.
- The responses from respondents that best represent their cluster. These respondents are extracted using two different criteria. *Paragons* are the respondents closest to the cluster's CoG. The *specific* respondents are those who are farthest from the other CoGs.

6.6 Other MFACT application types

MFACT can be used to evaluate products based on numerical ratings and free-text comments. This method is common in food preference studies, in which experts and/or consumers express their opinions through liking scores and comments. MFACT considers both types of evaluations when determining the distances between products, which provides a balanced contribution to the final score. More information can be found in Bécue-Bertaut et al. (2008). Furthermore, this approach is useful in socioeconomic surveys to address open-ended questions in addition to traditional questionnaires. This type of data is useful and widely accepted as a supplement to large questionnaires, and it is frequently collected. Thus, the proposed methodology can serve as a flexible marketing tool.

6.7 Results from the example

The results were obtained using the script **chap6**. The table in Section 3 is based on the Children and Reading Survey (see Section 2) and was analyzed using MFACT, followed by clustering. The analysis included only those who answered all of the questions. 795 people did so. Comments on the results are provided below.

6.7.1 MFACT

The reweighting of the columns in MFA balances the influence of the three sets when determining the first global axis. Set 3 (closed questions) contributes the most at 45.1%. This percentage is higher than the contributions of the two open-ended question sets, which are 25.7% and 28.3%, respectively. However, it is satisfactory that the closed question block contributes equally to the inertia of the first global axis as the open-ended question block. Similar values are found for the second axis. For the third axis, the influence of the open-ended questions is clearly predominant. This is consistent with the finding that the closed-question set presents two dominant first axes when analyzed separately using MCA.

MFACT inherits the fact that lexical tables derived from open-ended questions are large, sparse matrices. Consequently, the inertia of the first axis accounts for only a small fraction of the total variance. In this case, the first two MFACT axes retain 2.3% and 1.9% of the total inertia, respectively, with corresponding eigenvalues of 1.4 and 1.2. Though these percentages are small and similar to those observed when CA is performed on a lexical table obtained from open-ended questions, they are significant with respect to the hypothesis of independence between respondents and words. The high level of noise in respondents' responses does not prevent word structure from being revealed on the first factorial plane. Furthermore, these percentages should not be interpreted as percentages of information.

The first axis is a common dispersion direction among the three sets. However, it differs from the first axes of each separate analysis. This aligns with the first global analysis's eigenvalue of 1.4, which is far from the maximum value of 3 that would be obtained if the first axes of the separate analyses were identical.

Figures 6.1 and 6.2 display the respondents, categories, and word clouds from the MFACT analysis. Since the words and categories occupy the same space, overlaying the two clouds is appropriate. However, doing so would result in an overloaded map.

The outer envelope of the respondent cloud resembles a parabola. The interior is filled in. The upper right portion is stretched and scattered. A similar shape would be observed for the word cloud if all words were represented and the labels were removed. In contrast, the shape of the category cloud resembles a simplified parabola because each path connects only three points.

MFACT inherits the parabolic shape—also known as the Guttmann effect—from CA. This effect has already been discussed in previous chapters. The first axes show how the rows and columns are ordered with respect to each other. For the respondents (rows), this order is determined by the

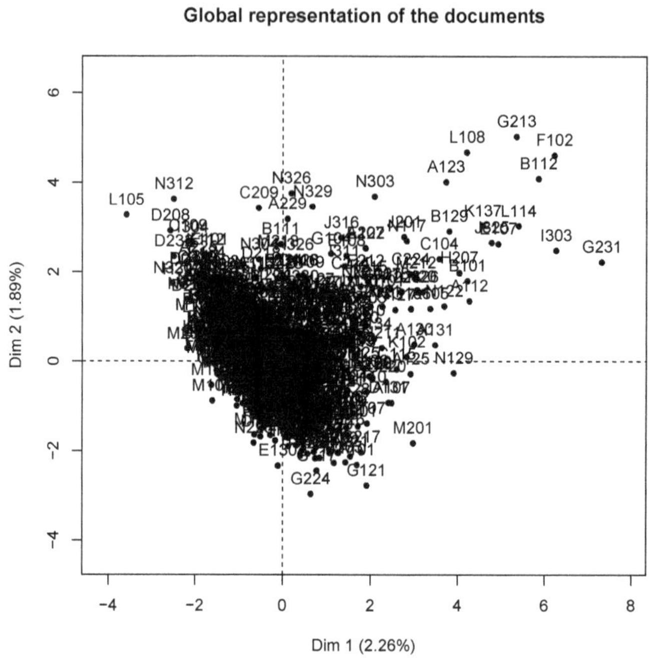

Figure 6.1: MFACT. Respondent cloud on the MFACT plane (1,2).

categories and words they select. For the categories and words (columns), the order is determined by the respondents who use them. The first axis ranks both the rows and columns according to this order. Meanwhile, the second axis contrasts elements with extreme coordinates on the first axis with those that have intermediate coordinates.

When analyzing open-ended questions with CA, the *filled parabola* pattern often emerges. A sharp parabola cannot result because verbal expression is inherently complex and sometimes contradictory. The mutual ordering of words and respondents is never as clear as an ordering that could emerge from the relationship between *age group* and *health status*, to give an example. Nevertheless, to better understand the obtained visualizations, the parabola form must be considered.

The words are arranged from those indicating a strong attraction to reading on the left to those reflecting a rejection of reading on the right. Words positioned far to the left include *learning, imagination, they learn* and *you have fun*, while words positioned far to the right include *boredom* and *drag.* As expected, the three categories on the far left are *I read a lot, I read easily* and *We have many books at home.* The three categories on the far right are *I read with great difficulty, We have few books at home* and *I read little.* Therefore, this axis can be interpreted as an axis of attraction to reading. From the

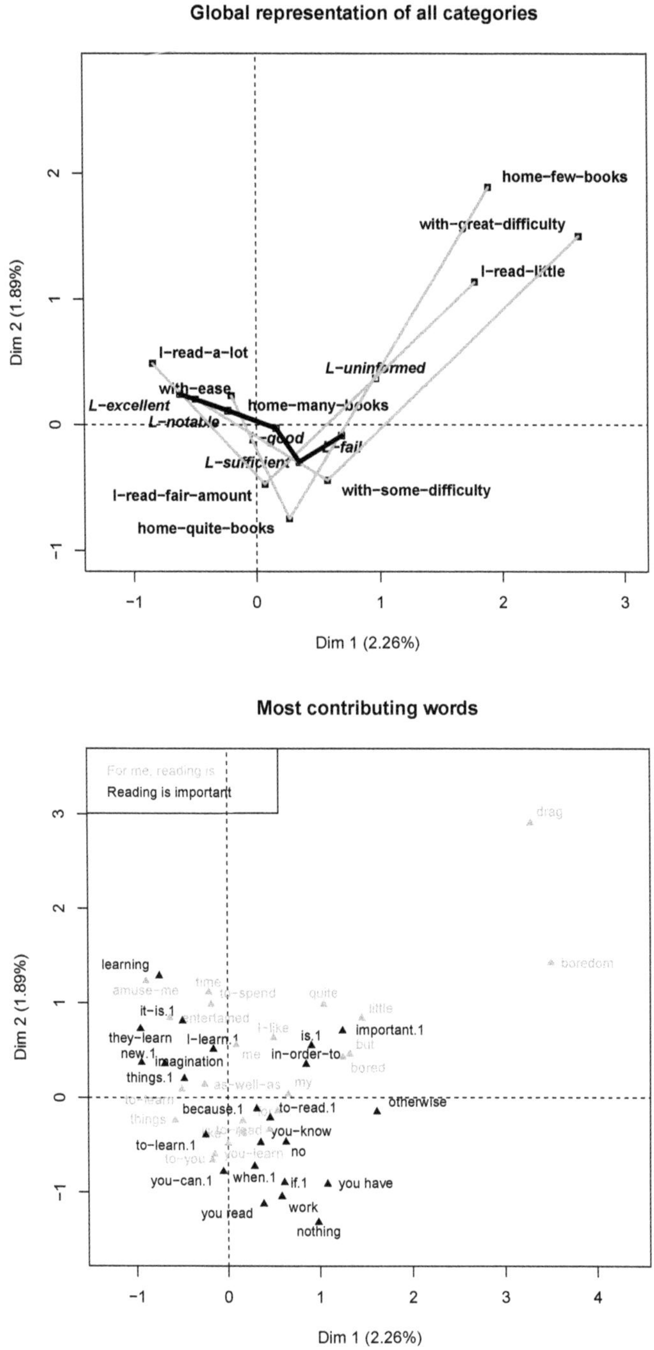

Figure 6.2: MFACT. Column cloud on the MFACT (1,2). Top: categories; bottom: words.

transition relationships, we can deduce that the respondents are ordered from good readers on the left to poor readers on the right.

This is the overall trend, though some respondents and words do not follow it. Nonconformist respondents are likely to use words from both ends of the first axis in their responses. These anti-trend words are likely used by respondents situated at opposite ends of the first axis. This phenomenon is partly explained by linguistic versatility.

The second axis contrasts good and poor readers with moderate readers. It also contrasts words indicating moderate attraction to reading with words indicating strong attraction or aversion. Note that respondents and words with high positive coordinates on the first two axes are widely dispersed. These elements accentuate the characteristics that typify this part of the map. The children are poor readers. Only negative words are found, such as *drag* and *boredom*.

Due to the open-ended questions, the third and fourth axes offer additional, albeit minor, insights into specific verbal usage. These axes will be used in the clustering step.

6.7.2 Clustering results

First, we perform a classical hierarchical ascending clustering analysis based on the first four factorial axes, as recommended by Cattell's scree test. This method is efficient and widely accepted. Together, these four axes explain 7.6% of the total inertia. As previously mentioned, the inertia rates corresponding to the first axes are necessarily weak in this type of analysis. The hierarchical tree (not shown here) suggests partitioning into four clusters, which we then follow. Next, we use this initial partition as a starting point to run a k-means algorithm to consolidate the clustering. The final partition obtained in this way has an interclass/total inertia ratio of 51.5%, indicating that the clusters have no clear boundaries. This is consistent with a fairly continuous respondent cloud without clearly bounded clusters (see Figure 6.1). The clusters are identified by a number according to the position of their CoG on the first axis, with number 1 being on the far left. In the following sections, we report only the most salient features of the clustering results.

6.7.2.1 Description of partition

The cluster sizes are 229, 250, 275, and 41, respectively. Three of the clusters have balanced numbers (clusters 1 through 3), while the fourth (cluster 4) is residual. Figure 6.3 shows the configuration of respondents colored according to their cluster on the first MFACT map. This figure combines the continuous view of the data provided by factor analysis with the essentially discontinuous view offered by clustering.

Table 6.5 shows the degree of association between the partition and the questions, whether closed- or open-ended, as determined by Cramer's V.

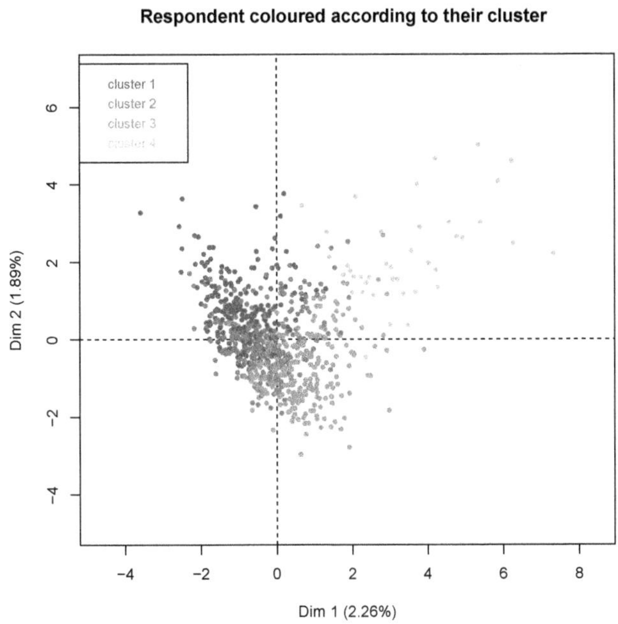

Figure 6.3: MFACT. Respondents on plane (1,2) colored from the cluster they belong to.

The question most strongly associated with the partition is the one regarding the amount of reading, with categories ranging from *I read very little* to *I read a lot*. Cramer's V for this question is 0.54. However, the two open-ended questions show a fairly close degree of association, with Cramer's V values of 0.48 and 0.45, respectively. The other questions are less closely associated with the partition, particularly the verbal skills score, which plays a supplementary role.

6.7.2.2 Description of clusters

Consider Clusters 1 and 4, for example. Based on their positions on the first axis, Cluster 1 contains the best readers, and Cluster 4 contains the worst. Cluster 1 is in the second quadrant and is concentrated along the second bisector. Cluster 4 is in the first quadrant and is the residual cluster. This cluster includes children ranging from true poor performers to those who perform slightly better.

Table 6.6 shows the overrepresented categories within the two clusters. How should this table be read? Let's use the first line as an example. It provides information about the *I read a lot* category of the categorical variable *amount of reading*. The `Cla/Mod` value of 73.64% indicates that 73.64% of

Table 6.5: Measure of the association between partitions and questions.

Question	*Cramer's coefficient*
I read (amount of reading; closed)	0.54
For me, to read means (open-ended)	0.48
Reading is important because (open-ended)	0.45
I read (ease of reading; closed)	0.40
At home we have (amount of books; closed)	0.38
Verbal skills score (closed)	0.33

Table 6.6: Description of clusters 1 and 4 in terms of the closed questions.

```
Cluster 1 - overrepresented categories
                                 Cla/Mod Mod/Cla Global p.value  v.test
I read=a lot                      73.636  70.742 27.673   0.000  16.887
I read with=ease                  42.950  86.463 57.987   0.000  10.834
at home we have=many books        35.413  84.279 68.553   0.000   6.299
Verbal score: L-excelente         41.525  21.397 14.843   0.001   3.206
Verbal score: L-notable           35.088  34.934 28.679   0.014   2.449

Cluster 4 - overrepresented categories
I read=little                     30.631  82.927 13.962   0.000 10.304
I read with=many difficulties     57.692  36.585  3.270   0.000  7.676
At home we have=few books         43.333  31.707  3.774   0.000  6.388
Verbal score: L-fail              16.327  19.512  6.164   0.003  2.988
```

respondents who reported reading a lot belong to this cluster. The `Mod/Cla` value of 70.74% shows that 70.74% of the cluster's respondents selected *I read a lot*. The `Global` value of 27.67% shows that 27.67% of all respondents chose the *I read a lot* category. A statistical test (see Husson et al., 2009, Chap. 4) compares the `Mod/Cla` and `Global` proportions. This calculation yields a `p-value` of 0.000. The `p-value` is converted to a `v-test` of 16.89, which is the quantile of the normal distribution associated with the `p-value`. The sign of the `v-test` indicates whether a category is overrepresented (positive sign) or underrepresented (negative sign). Here, we only show overrepresented categories.

Table 6.7 lists the characteristic words of each cluster. See the first line for an example of how to read the table. The first line corresponds to the word *amuse-me*. `Intern` = 1.56% means that *amuse-me* corresponds to 1.56% of the responses from Cluster 1 respondents. However, this word corresponds to only 0.60% of the corpus size (=all responses from all respondents). `Freq` = 29 means that *amuse-me* is repeated 29 times in Cluster 1 responses, whereas it is used 33 times in all respondents' responses. The proportions `Intern` and `Glob` are compared using a statistical test that yields a `p-value` of 0.000. This

Table 6.7: Description of clusters 1 and 4 in terms of its most overrepresented characteristic words.

Cluster 1	overrepresented words - tools words are released					
	Intern %	glob %	Intern freq	Glob freq	p.value	v.test
amuse-me	1.559	0.604	29	33	0.000	6.355
to-spend	1.129	0.512	21	28	0.000	4.370
adventures	1.237	0.659	23	36	0.000	3.621
moment	1.022	0.512	19	28	0.000	3.601
amusement	1.344	0.769	25	42	0.001	3.353
time	0.645	0.293	12	16	0.001	3.261
I.have-fun	0.430	0.183	8	10	0.004	2.852
you-have-fun	0.430	0.220	8	12	0.026	2.230
studying	0.699	0.421	13	23	0.030	2.176
to-know	0.430	0.238	8	13	0.049	1.967
I-spend	0.430	0.238	8	13	0.049	1.967

Cluster 4	overrepresented words - tools words are released					
	Intern %	glob %	Intern freq	Glob freq	p.value	v.test
drag	5.446	0.201	11	11	0.000	8.269
boresome	3.465	0.183	7	10	0.000	5.741
like	2.970	0.988	6	54	0.017	2.378
(in the expression do not like)						

`p-value` is converted to a `v-test` of 6.36. Once again, the sign of the `v-test` indicates overrepresentation or underrepresentation.

Table 6.8 shows successively the three responses of Clusters 1 and 4 that best meet each of the two criteria. The first criterion identifies respondents who are closest to their CoG. These respondents are called *paragons*. The second criterion selects respondents who are farthest from the CoGs of the other clusters. These respondents are called *specific*. A comprehensive analysis of these responses provides context for the clusters' characteristic words.

Similarly, we extracted characteristic words and responses from the second open-ended question. These results are not shown here. A synthetic reading of all the results provides insight into the characteristics of each cluster of respondents. A few words about that: Cluster 1 consists of high-achieving students, most of whom *read a lot* (71% versus 28% of the full sample). Note that children are assigned to this cluster based on their responses to both closed- and open-ended questions. In their free responses, they express a positive attitude toward reading, viewing it as enjoyable and as an opportunity for adventure. Their responses reveal a vocabulary beyond their years. Unsurprisingly, 56% received an outstanding or notable score for verbal skills in school. Cluster 4 is completely different. It contains students who struggle with reading and rarely practice. Their responses show little enthusiasm for reading; in fact, they reject it altogether, considering it *a drag*. Their poor performance in this subject is consistent with their awkward and inadequate verbal expression.

Table 6.8: Most characteristic responses to "for me, to read means". Clusters 1 and 4.

```
Cluster 1. Responses of the three first paragons
[1] when I have nothing to do, I pick up a book and have fun reading adventures
    and mysteries.
[2] learn the letters so you can find out what they say in books, newspapers,
    etc.
[3] have fun, live adventures, and learn

Cluster 1. Responses of the three first specific respondents
[1] amuse-me
[2] amuse-me
[3] to learn and amuse me etc


Cluster 4. Responses of the three first paragons
[1] fun
[2] Something very fun, if they are adventurous, I like it.
[3] important to my life health

Cluster 4. Responses of the three first specific respondents
[1] quite a drag
[2] a drag
[3] a drag
```

6.8 Conclusion

Treating a mixture of data consisting of sets of quantitative, categorical, and frequency columns simultaneously is useful for a wide range of applications. MFACT, an extended multiple factor analysis (MFA) method, facilitates this process. MFACT computes a global distance between rows, balancing the influence of the column sets and enabling respondents to be clustered based on all of them. This technique is best applied to survey data because users naturally consider the responses to open- and closed-ended questions when clustering questionnaire respondents. Closed questions provide a solid framework, while open-ended questions contribute diversity, which can be confusing when considered separately. Considering both types of questions simultaneously combines their respective advantages.

7

Correspondence analysis on a generalized aggregate lexical table

7.1 Introduction

Chapter 2 presented CA applied to a DTM (also known as a lexical table). When analyzing open-ended responses from questionnaire-based surveys, the structural weaknesses of the table become apparent when respondents provide brief responses. In such cases, the profiles and distances between respondents exhibit significant variability. This results in axes that primarily reflect local phenomena, hindering the identification and synthesis of critical information.

It is common practice to group documents according to descriptive categories such as *age group* or *education level.* Then, CA is applied to the aggregated lexical table (ALT), which crosses categories (rows) and words (columns). This method yields more robust results because it controls for respondent characteristics (Lebart et al., 1998, p. 103–104). CA on the ALT (CA-ALT) provides a symmetric approach to the relationships between words and categories. It allows the variability observed between words to be explained by the variability observed between categories, and vice versa. Attractions and rejections between particular words and categories are identified and visualized on the CA principal planes. In a sense, the grouping variable plays an explanatory role with respect to word choice.

Considering several categorical variables would provide a better understanding of the variability observed in lexical choices. However, it is not practical to group respondents by categories formed by combining all the variables' categories because there are too many. Additionally, due to the potential complexity of the relationships between the variables, some categories would be empty or have low counts. Another approach would be to apply CA to the table resulting from juxtaposing the aggregated lexical tables built from each categorical variable. The main drawback of this approach is that it does not eliminate associations between variables. Therefore, confounding effects may remain and obscure the influence of each variable or category on vocabulary choice. This chapter presents a methodology that helps avoid this problem. Furthermore, this methodology can be applied when the explanatory variables are quantitative, thus broadening its scope. The analyzed table is constructed from textual and contextual tables using logic similar to that used to create an

DOI: 10.1201/9781003397472-7

ALT. It is called a *generalized aggregated lexical table* (GALT). The methodology to deal with this table is called CA on a GALT (CA-GALT). It is a form of linearly constrained CA (Takane et al., 1991), specifically a special case of canonical CA (ter Braak, 1986, 1987).

This chapter is organized as follows: Section 7.2 provides a motivating example. Section 7.3 introduces the notation. Section 7.4 reviews the traditional approach to an aggregated lexical table. Section 7.5 describes CA-GALT, a suitable method for analyzing a GALT. Section 7.6 evaluates its effectiveness using the example. Section 7.7 concludes with some remarks.

7.2 Example: Health

This survey has been used extensively in previous chapters. Here are a few words to serve as a reminder. The objective is to determine laypeople's opinions and attitude toward health. The study examined 513 people over the age of 14. The first question of the questionnaire, *What does health mean to you?*, required free and spontaneous responses. The researchers considered the following possible conditioning variables of the respondents' view of health: *age group* (*under 21, 21:35, 36:50*, and *over 50*); *gender* and *health status* (*bad, fair, good*, and *very good*). The main objective was to identify different health-related concerns and their relationship with the respondents' characteristics. Here, we use the translated into English and lemmatized free responses.

7.3 Notation

Matrices $\mathbf{X}$ and $\mathbf{Y}$ are constructed from the responses of I respondents (see Figure 7.1). Matrix $\mathbf{X}$ is an $I \times K$ matrix that stores K contextual variables, which can be either quantitative or categorical but all of the same type of variable. The generic term for matrix $\mathbf{X}$ is x_{ik}. The categories are coded as dummy variables. $\mathbf{Y}$ is an $I \times J$ matrix containing the frequency of J words used in the free responses of I respondents. The generic term for matrix $\mathbf{Y}$ is y_{ij}. In total, N occurrences of J words are observed.

The proportion matrix $\mathbf{P}$ is calculated as $\mathbf{P}=\mathbf{Y}/N$. Its generic term is $p_{ij} = y_{ij}/N$. The column and row margins of $\mathbf{P}$ are stored in the $(I \times I)$ diagonal matrix $\mathbf{D_I}$ and the $(J \times J)$ diagonal matrix $\mathbf{D_J}$, respectively. The generic terms are $p_{i.} = \sum_j p_{ij}$ and $p_{.j} = \sum_i p_{ij}$. From $\mathbf{P}$, the double standardized $(I \times J)$ matrix $\mathcal{T}$ is defined as $\mathcal{T}=\mathbf{D_I^{-1}PD_J^{-1}}$. Its generic term is $\tau_{ij} = p_{ij}/(p_{i.}p_{.j})$. This generic term is equal to the ratio of the observed proportion (p_{ij}) to the expected proportion $(p_{i.}p_{.j})$ according to the independence

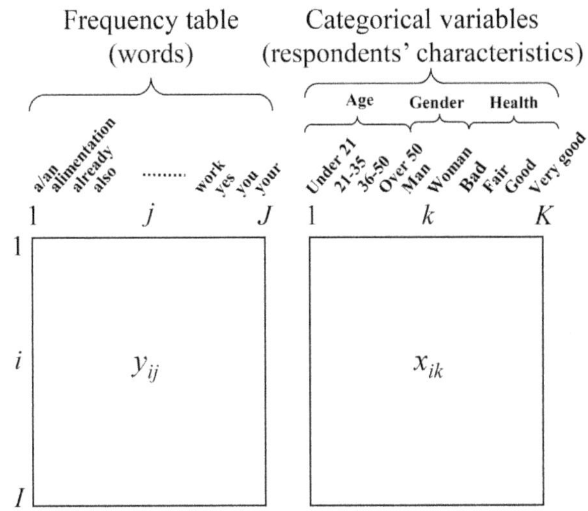

Figure 7.1: The data set. On the left, the frequency table **Y**; on the right, the categorical table **X**. There are I respondents, J words, and K categories.

model. This ratio measures the association between respondent i and word j. Matrix $\mathcal{T}$ is called the *association rate table* between respondents and words (Benzécri, 1981). This matrix was previously presented in Chapter 2. Another form of this matrix is the doubly centered form, denoted by the symbol $\bar{\mathcal{T}}$ with the generic term $\bar{\tau}_{ij} = (p_{ij} - p_{i.}p_{.j})/(p_{i.}p_{.j})$. These matrices have already been defined in Chapter 2.

7.4 One explicative categorical variable

In chapter 2, we learned that applying CA to **Y** involves calculating GSVD($\mathcal{T}$, $\mathbf{D_J}$, $\mathbf{D_I}$). This places the method within the general framework of principal axis methods. Alternatively, GSVD ($\bar{\mathcal{T}}$, $\mathbf{D_J}$, $\mathbf{D_I}$) can also be used. This second formulation emphasizes that CA analyzes the weighted deviation between **P** and the $(I \times J)$ independence model matrix $[p_{i.}p_{.j}]$.

Note that CA's analysis of an ALT—which groups individual responses belonging to the same category of a contextual variable into a single row—makes it possible to explain lexical choices as a function of this variable.

Let **X** be the $(I \times K)$ contextual table, whose columns are dummy variables corresponding to the categories of a single categorical variable. First, the $(J \times K)$ ALT matrix $\mathbf{Y_A} = \mathbf{Y^T X}$ is constructed by crossing the words and the categories of the categorical variable. Then, the $(J \times K)$ proportion matrix is calculated as:

$$\mathbf{P_A} = \frac{\mathbf{Y_A}}{N} = \mathbf{P^T X}.$$

The $(J \times J)$ diagonal matrix $\mathbf{D_J} = [d_{Jjj}] = [p_{A_j.}] = [p_{.j}]$ and the $(K \times K)$ diagonal matrix $\mathbf{D_K} = [d_{Kkk}] = [p_{A.k}]$ store the column and row margins of $\mathbf{P_A}$. $\mathbf{D_J}$ and $\mathbf{D_K}$ corresponds to the weighting system on the rows and the columns, respectively.

The double standardization of the matrix $\mathbf{P_A}$ is calculated as follows:

$$\mathcal{T}_\mathbf{A} = \mathbf{D_J^{-1} P_A D_K^{-1}}.$$

Then, CA-ALT, that is CA($\mathbf{Y_A}$), is implemented via GSVD($\mathcal{T}_\mathbf{A}, \mathbf{D_K}, \mathbf{D_J}$).

Remarks.

- $\mathbf{D_K} = \mathbf{X^T D_I X}$ is diagonal because $\mathbf{X}$ is a column-wise $\mathbf{D_I}$-orthogonal matrix. $\mathbf{D_K}$ acts as a covariance matrix when the analysis is performed on the non-centered matrix.
- The $\mathbf{X D_K^{-1/2}}$ columns are standardized and their inertia with respect to the origin is equal to 1.
- CA-ALT, i.e. GSVD($\mathcal{T}_\mathbf{A}, \mathbf{D_K}, \mathbf{D_J}$), is a double-projected analysis. This analysis explains the variability of the rows in terms of the columns and vice versa. It provides robust results indicating the associations (or oppositions) between words in relation to identical or different categories of the contextual variable.

7.5 Principles of CA-GALT

Using the contextual matrix $\mathbf{X}$, which contains K quantitative or dummy variables, a broader context is considered. If the variables are quantitative, they are centered and may be reduced to a unity variance. If the variables are dummy variables, then they correspond to the K categories of the categorical contextual variables. The GALT is calculated as the $(J \times K)$ matrix $\mathbf{Y_A} = \mathbf{Y^T X}$.

We use the same procedure to analyze this table as in the previous section. First, $\mathbf{Y_A}$ is transformed into the $(J \times K)$ matrix $\mathbf{P_A}$:

$$\mathbf{P_A} = \frac{\mathbf{Y_A}}{N} = \mathbf{P^T X}.$$

Its generic term $p_{A_{jk}} = \sum_i p_{ij} x_{ik}$, is the weighted sum of the values that respondents who used word j assumed for variable k.

The diagonal matrix $\mathbf{D}_K$, which stores the row margin of $\mathbf{P}_A$, is no longer equal to the $(K \times K)$ covariance matrix $\mathbf{C} = (\mathbf{X}^\mathbf{T}\mathbf{D}_\mathbf{I}\mathbf{X})$. Therefore, to obtain a doubly projected analysis, $\mathbf{D}_K$ must be replaced by $\mathbf{C}$. If $\mathbf{C}$ is not invertible, the Moore-Penrose pseudoinverse $\mathbf{C}^-$ is used to calculate the double standardization of the $(J \times K)$ matrix $\mathbf{P}_A$:

$$\mathcal{T}_\mathbf{A} = \mathbf{D}_\mathbf{J}^{-1}\mathbf{P}_\mathbf{A}\mathbf{C}^-.$$

Finally, CA-GALT involves calculating the GSVD $(\mathcal{T}_\mathbf{A}, \mathbf{C}, \mathbf{D}_\mathbf{J})$ (Bécue-Bertaut et al., 2014b; Bécue-Bertaut and Pagès, 2015; Kostov et al., 2015). CA-GALT provides standard GSVD results. The singular values are stored in the $(S \times S)$ diagonal matrix $\mathbf{\Sigma}$. The left and right singular vectors in the $(J \times S)$ matrix $\mathbf{U}$ and the $(K \times S)$ matrix $\mathbf{V}$, respectively. The coordinates of the row words are computed as follows:

$$\mathbf{F} = \mathcal{T}_\mathbf{A}\mathbf{C}\mathbf{V}.$$

The coordinates of the column variables are computed as follows:

$$\mathbf{G} = \mathcal{T}_\mathbf{A}^\mathbf{T}\mathbf{D}_\mathbf{J}\mathbf{U}.$$

Remarks

1. Interestingly, the $(J \times K)$ table $\mathbf{D}_\mathbf{J}^{-1}\mathbf{P}_\mathbf{A}$ has the generic term:

$$\bar{x}_{kj} = \sum_{i=1}^{I} \frac{p_{ij}}{p_{\cdot j}} x_{ik}.$$

   In other words, the value of each word j for variable k is determined by weighted averaging.

2. $\mathbf{C}^-$ performs multivariate standardization on the columns of $\mathbf{X}$ but also renders them uncorrelated (Brandimarte, 2011; Härdle and L.Simar, 2012). The contextual variables $\mathbf{X}$ are now considered through the subspace they span.
3. Applying the GSVD$(\mathcal{T}_\mathbf{A}, \mathbf{C}, \mathbf{D}_\mathbf{J})$ is equivalent to applying the CA to the $\mathbf{P}_A$ with a *modified metric* in the row space.
4. As in CA, the singular values lie between 0 and 1.
5. The word and variable factors are linked by the standard transition relationships. Thus, the word factors stored in $\mathbf{F}$ are expressed in terms of the variable factors stored in $\mathbf{G}$ as follows:

$$\mathbf{F} = \mathcal{T}_\mathbf{A}\mathbf{C}\mathbf{G}\mathbf{\Sigma}^{-1} = \mathbf{D}_\mathbf{J}^{-1}\mathbf{P}^\mathbf{T}\mathbf{X}\mathbf{G}\mathbf{\Sigma}^{-1}.$$

   Similarly, the factors on the variables are expressed as a function of the factors on the words as follows:

$$\mathbf{G} = \mathcal{T}_\mathbf{A}^\mathbf{T}\mathbf{D}_\mathbf{J}\mathbf{F}\mathbf{\Sigma}^{-1} = \mathbf{C}^{-1}\mathbf{X}^\mathbf{T}\mathbf{P}\mathbf{F}\mathbf{\Sigma}^{-1}.$$

The coordinates of the variables on the axes are equal to the covariances between the columns of $\mathcal{T}_A$ and the standardized factors on the words $\mathbf{F\Sigma}^{-1}$.

6. The interpretation rules are very similar to those of CA. In CA-GALT, two words are close if respondents with similar profiles in $\mathbf{X}$ use them. Two contextual categories or variables are close if respondents with these categories or with high values for these variables use the same words.
7. Another way to view the analysis scheme is to show that it combines regression and CA. First, each column of $\mathcal{T}$ is regressed on the contextual variables, with respondents weighted by $\mathbf{D_I}$. Then, the regression coefficients are calculated and stored in the $(K \times J)$ matrix:
$$\mathbf{B} = (\mathbf{X}^{\mathbf{T}}\mathbf{D_I}\mathbf{X})^{-1}\mathbf{X}^{\mathbf{T}}\mathbf{D_I}\mathcal{T}.$$
Next, the $(I \times J)$ matrix $\hat{\mathcal{T}}$ is obtained as
$$\hat{\mathcal{T}} = \mathbf{XB}.$$
Finally, GSVD$(\hat{\mathcal{T}}, \mathbf{D_J}, \mathbf{D_I})$ is performed. This emphasizes the regression aspect of the method, indicating that it only analyzes the portion of word usage that is explained by contextual variables.
8. In classical principal axes methods, the diagonal matrix serves two functions: it weights the elements in one space and acts as a metric in the other space. However, in the case of a non-diagonal metric matrix, such as in word space, the weighting function is lost, making the transition relationships unclear.
9. Respondents can be reintroduced into the analysis by adding the columns of $\mathcal{T}^T$ as supplementary columns in GSVD$(\mathcal{T}_\mathbf{A}, \mathbf{C}, \mathbf{D_J})$. Thus, their coordinates are computed via the transition relationships.

7.6 Results from the example

We use the data presented in Section 2. The results were obtained using the script **chap7**.

7.6.1 Pre-processing of the data

CA-GALT is applied to the DTM obtained from this version of the free responses (textual data, matrix $\mathbf{Y}$) and the three contextual categorical variables that are *age group* (four categories), *gender* (two categories) and *health status* (four categories) (contextual data, matrix **X**). The lemmas *work_NOUN* and *job_NOUN* are considered equivalent since they correspond to the same Spanish word, *trabajo*. A total of 392 respondents who answered both open-ended and closed questions were selected for the study. Only lemmas used at least ten times are kept. The final corpus consists of 8,540 occurrences of 140 different lemmas.

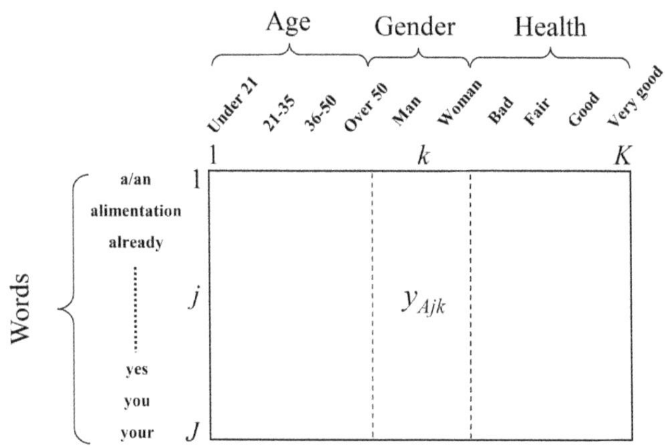

Figure 7.2: Generalized aggregated lexical table with the words in rows and the categories of age, gender and health status in columns.

7.6.2 CA-GALT on the multiple aggregated table

CA-GALT is applied to the multiple table $(\mathbf{X}, \mathbf{Y})$. The maximum number of non-zero eigenvalues, seven, was obtained. This indicates low redundancy between the categories of the different contextual variables. The total inertia is 0.18, which is equal to the sum of all the eigenvalues or squared singular values. The first axis is clearly dominant. It is associated with an eigenvalue of 0.060, which accounts for 32.69% of the total inertia. The next two axes stand out slightly, with rounded eigenvalues that are both equal to 0.026. These axes account for 14.27% and 14.00% of the inertia, respectively. The fourth axis is associated with an eigenvalue of 0.022, accounting for 12.08% of the inertia.

Figures 7.3 and 7.4 illustrate the most prevalent words and contextual variables on planes (1,2) and (1,4), respectively. The trajectories of the categories are shown by *age group* and *health status*. These representations are supplemented by plotting confidence ellipses (Efron, 1979; Lebart et al., 2006). Ellipses are drawn around all categories. Only the ellipses around the lemmas *the_DET*, *best_ADJ*, *main_ADJ*, on the plane (1,2), and *work_NOUN*, on the plane (1,4), are shown, because displaying all of them would be confusing. These words were chosen to demonstrate the effectiveness of the proposed approach.

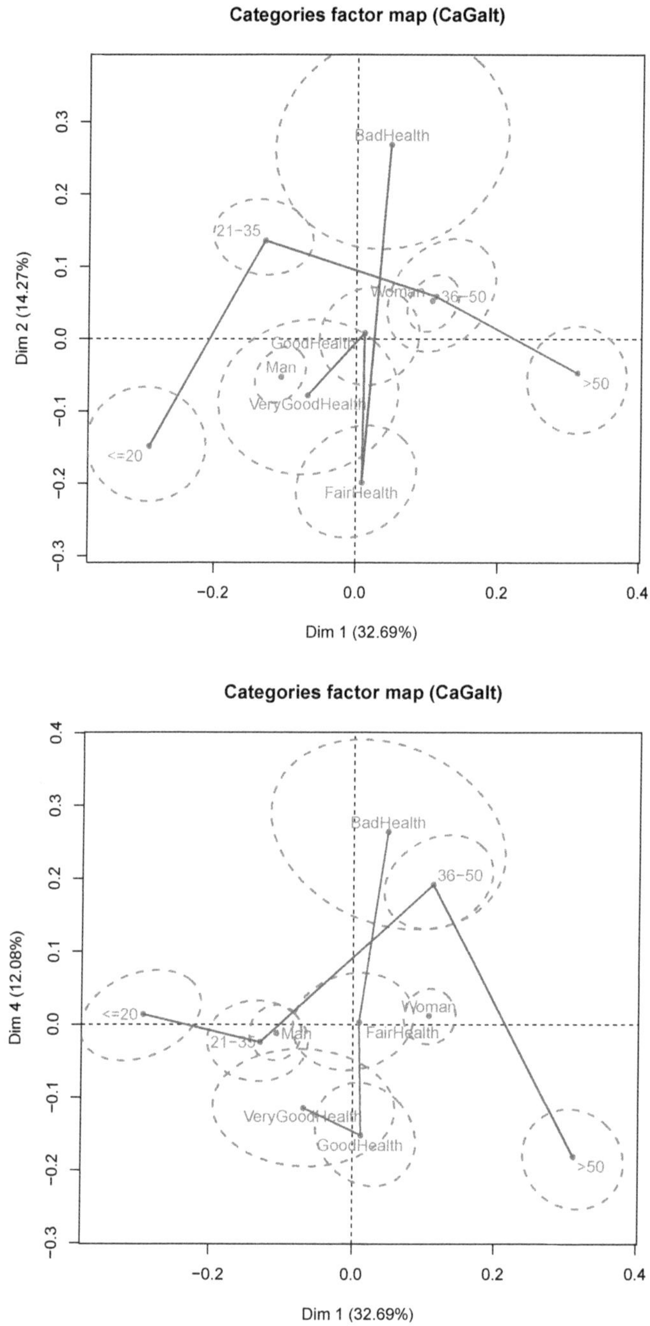

Figure 7.3: Categories on CA-GALT planes completed by confidence ellipses. Top: plane (1,2); bottom: plane (1,4).

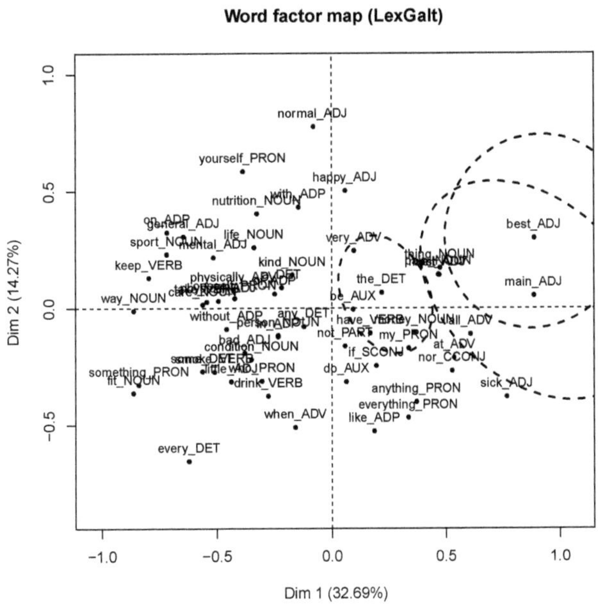

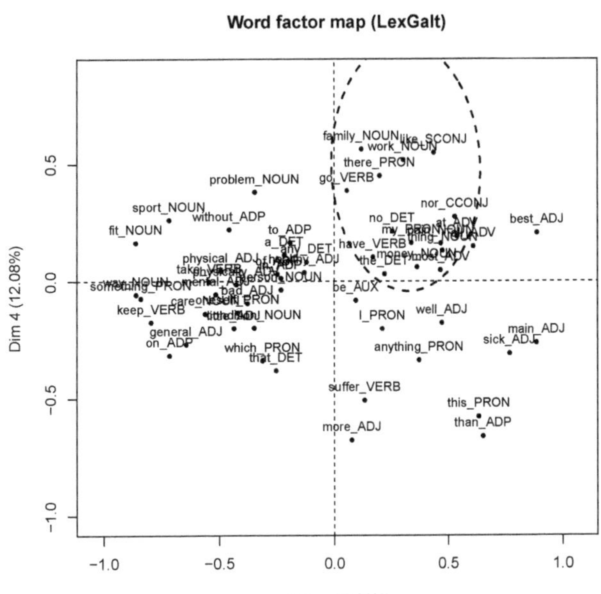

Figure 7.4: Best represented words on CA-GALT planes completed by some confidence ellipses. Top: plane (1,2); bottom: plane (1,4).

Table 7.1: Cross table Age group × Health status.

	Health status				
Age group	Bad	Fair	Good	V Good	margin
≤ 20	0	19	44	8	71
21 – 35	2	37	81	15	135
36 – 50	2	34	33	7	76
> 50	21	49	36	4	110
margin	25	139	194	34	392

On the plane (1,2), the variable *age group* follows a parabolic trajectory, corresponding to the Guttman effect (see Figure 7.3, top). This effect means that the first axis arranges categories and words in an orderly fashion relative to each other. Meanwhile, the second axis contrasts points with extreme coordinate values on the first axis with points that have intermediate values on the same axis. This allows us to observe primary changes in vocabulary associated with age. The categories appear in their natural order because word choice is age-related. The order of the words is determined by the categories of the respondents who use them. This enables us to identify the words used by young, middle-aged, and elderly people, respectively, in response to the open-ended question. The Guttman effect is less pronounced for words than for categories because the order of words affects those related to topics strongly associated with the question, but not other words, particularly tool words.

For example, the words *the_DET*, *best_ADJ*, and *main_ADJ* appear to the right of the first axis. This indicates that these words are frequently used by the oldest categories and avoided by the youngest. The word *work_NOUN* is absent from this graph because it does not appear frequently enough in the first plane. Therefore, it is not a key word for the oldest or youngest categories, and its use is unrelated to age.

The third axis contrasts *man* (with a positive coordinate) and *woman* (with a negative coordinate). This axis is not shown here. Thus, it also contrasts masculine words, such as *sport_NOUN* and *smoke_VERB*, with feminine words, such as *pain_NOUN*. This is confirmed by applying the test discussed in Chapter 6, which compares a word's frequency in a subcorpus to its frequency in the entire corpus to detect overuse or underuse. The results show that *sport_NOUN* and *smoke_VERB* are significantly overused by men (`p-value` equal to 0.009 and 0.082, respectively). Conversely, *pain_NOUN* is overused by women (`p-value` equal to 0.004).

The fourth axis is interesting because it ranks the *health status* categories in their natural order, from worst to best, except for the two categories of best health status. As their overlapping confidence ellipses indicate, these categories *Very good health* and *Good health* are virtually identical in terms of the associated vocabulary (see Figure 7.3, bottom). We chose to reproduce plane (1,4) rather than (3,4) because the former clearly illustrates the relationship between *health status* and *age trajectories*. These two trajectories unfold along axes 1 and 4, respectively, demonstrating that the influence of these variables on vocabulary choices is uncorrelated.

Bad health stands out at the top of the axis, very clearly separated from *fair health*, which is also clearly separated from the best categories. Examining the words, we see that *work_NOUN* is close to the *bad health* category in the positive part of the fourth axis (see Figure 7.4, bottom). This indicates a strong association between the word and the category, as well as low usage by the healthiest categories. On the fourth axis, the word *work_NOUN* contrasts with *more_ADJ*, *suffer_VERB* and *anything_PRON* which are associated with *good health*, *very good health* and *over 50*. These words are often used in negative expressions, such as *not to suffer*, *more than anything*, *not to suffer from anything* or *from pain*.

CA-GALT reveals a distinction between words associated with *bad health* and those associated with *over 50*. This finding must be verified using the data. A variable crossing *age group* and *health status* is created. Similar categories are grouped together to ensure that each contains a minimum number of members. This cross-variable allows us to compare vocabulary from the perspectives of *health status* at the same *age group* and vice versa. We can determine if words are significantly under- or overused than expected for each category by applying the aforementioned test. We keep not only the words that are significantly under- or overused (`p-value` < 0.1) but also the words that are moderately under- or overused ($0.1 <$ `p-value` < 0.16) because the progression of a word's use depending on increasing age and decreasing *health status* is also of interest. The results corresponding to the four words (*the_DET*, *best_ADJ*, *main_ADJ* and *work_NOUN*) are summarized in Table 7.2.

As shown in Table 7.2, respondents who rarely use the words *the*, *best* and *main* differ from those who overuse these words in expressions such as *the best* or *the main thing*. Only respondents *over 50* with significantly different *health statuses* overuse these words to a moderate or significant extent. The usage of these three words depends on *age group* rather than *health status* and increases with *age*.

We will now focus on the word *work_NOUN*. It is underused by those aged *21-35* with *good health* and by those aged ≤ 20 with *good* or *very good health*. It is significantly overused by those *over 50* with *bad health* and by those aged *36-50* with *bad/fair health*—the least healthy category in this age group. These results are less obvious than the previous ones, suggesting that *health status* alone does not affect the selection or rejection of this word. In this case, it is more difficult to disentangle the influence of *age* and *health status* because the *bad health* category consists almost exclusively of respondents *over 50* (21 out of 25). Nevertheless, those aged *over 50* and *36-50* with better health do not overuse *work_NOUN* even moderately. This allows us to conclude that *health status* has a stronger effect than *age* on the choice of this word.

CA-GALT performs a regression-like analysis to determine the influence of each variable. This analysis highlights the specific effect of each selected explanatory variable.

Table 7.2: Categories under- or over-using *the, best, main*, and *work*.

word	Significant under-use	Moderate under-use	Moderate over-use	Significant over-use
the_DET	$\leq$ 20 good/very good health 21 – 35 fair/bad health		$\leq$ 20 fair health	> 50 fair health > 50 bad health
best_ADJ	$\leq$ 20 good/very good health 21 – 35 fair/bad health		> 50 good/very good health	> 50 good/very good health
main_ADJ	21 – 35 fair/bad health 21 – 35 good health			> 50 good/very good health > 50 bad health > 50 fair health
work_NOUN		$\leq$ 20 good/very good health 21 – 35 very good health		> 50 bad health 36 – 50 bad/fair health

7.7 Conclusion

To identify the complex relationships between respondent characteristics and lexical choices, contextual variables are incorporated and assigned an explanatory role. This results in the creation of a GALT, which is then analyzed by CA-GALT using a strategy similar to CA's double-projected approach. This allows us to study vocabulary diversity through variable dispersion and variable dispersion through vocabulary diversity. Thus, the associations and/or oppositions between words derive their meaning from the variables they attract or reject, and vice versa. Applying this method to real data reveals how open- and closed-ended responses converge to provide meaningful insights. To avoid confounding effects, we visualize the influence of each variable on lexical choices. In the example, the words favored by the different categories reveal the health-related concerns associated with each variable (e.g., *age*, *health status*, or *gender*) in a ceteris paribus analysis. This approach clearly shows that the meaning of any discourse can only be understood when the speaker's characteristics are known. Rightfully, we can expect meaningful associations and/or oppositions between words.

8

Structure and organization of a text

This chapter is the work of Mónica Bécue-Bertaut, Belchin Kostov, Annie Morin and Guilhem Naro

8.1 Introduction

Textual statistics can greatly contribute to the study of discursive strategies. In particular, legal, rhetorical, and argumentative texts have interesting statistical characteristics. The main objective is to reveal the text's structure, which is crucial for convincing the audience. According to a renowned lawyer, authors of this type of text must *guide the reader to the only possible conclusion by presenting an irrefutable argument*. In other words, the text should appear as reality itself—the truth. To be persuasive, one must appeal to logical reasoning that effectively connects and supports arguments. This results in the regular renewal of vocabulary, which speeds up or slows down based on significant events or when the speaker wants to emphasize a particular point. To this end, we propose a methodology combining CA visualization tools, time-constrained hierarchical clustering, and topic extraction. CA is an essential text visualization tool because it tracks the trajectory of its constituent parts. Topic extraction identifies the words that characterize the trajectory and hierarchy.

The nature of the results obtained by applying this methodology is illustrated by its use in analyzing a closing argument issued by a prosecutor during a murder trial at the Barcelona Court of Justice. This closing argument is notable because there was no direct evidence linking the defendant to the crime. In such cases, prosecutors must rely on *circumstantial evidence*, which is used in legal proceedings to suggest the existence of a fact or event without confirming it directly. This requires prosecutors to work harder to build stronger cases.

The chapter is organized as follows: Section 8.2 provides an example, which is a closing argument delivered before a court. Section 8.3 introduces the methodology. Section 8.4 presents the results of the example and explains how to interpret them. Section 8.5 offers concluding remarks.

DOI: 10.1201/9781003397472-8

8.2 Example: a closing argument for the prosecution

JPM, the owner of a *sauna-relax* establishment where prostitution was practiced, and his accomplice, FPM, have been accused of planning and carrying out the murder of MJAM, a woman who worked at their establishment. The two men wanted to collect on a substantial life insurance policy purchased by MJAM. In fact, the sauna owner and his children, not MJAM's children, were listed as the beneficiaries of the insurance policy.

The trial lasted two days. At the end, the prosecutor spent an hour and fifteen minutes delivering the closing argument. The lack of material evidence complicated the prosecutor's role, resulting in a lengthy and detailed closing argument. This type of speech must be based on facts and supported by evidence that logically links them to the crime. As the trial progresses, the prosecutor drafts the closing argument based on the testimony of the defendants, witnesses, and experts. However, the prosecutor can only finalize the text at the end of the trial because its content depends on the outcome. It is not possible to write it out completely beforehand.

The structure is fairly standard. First, the prosecutor lists the charges against the defendant(s). Next, the applicable legal framework is stated, followed by a description of the facts that fit within it. Finally, the text concludes with a legal qualification of the facts.

However, each closing argument stems from the prosecutor's specific strategy based on their habits and style as well as the peculiarities and difficulties of the trial at hand. In this case, the prosecutor accuses the two defendants of three crimes. In addition to murder, the prosecutor alleges that they committed pimping and insurance fraud.

During his interrogation, the primary defendant claimed that the murdered woman was his romantic partner and that they had taken out life insurance policies on each other. He provided evidence showing that he had been out of town on the day of the murder and that he searched for her after returning, alarmed by her disappearance. The prosecutor attempted to refute this account and demonstrate its implausibility. According to the prosecutor, the defendant targeted a poor woman, convinced her that they were in a romantic relationship, and pressured her into taking out a life insurance policy. The prosecutor alleged that the defendant waited for the opportune moment to kill the victim so that he would have an alibi and be able to collect the insurance money. What is the basis for this accusation? There is no hard evidence. The only direct witness was a sauna employee who overheard the primary defendant and his accomplice discussing the plan to kill the victim. However, she retracted her initial statement to the police before the examining magistrate. The prosecutor emphasized how absurd the defendant's statements were regarding his relationship with the victim, as well as his actions on the day of the murder and during the supposed search for the victim.

The goal is to analyze how the prosecutor organizes and uses the available information. First, CA will outline the structure of the closing argument. Then, time-constrained clustering will divide the information into *periods*. Finally, characterizing these periods by their lexical features will lend meaning to the temporal structure.

8.3 Methodology

Our main goal is to reveal the structure of an argumentative text on two levels: its temporal evolution and the hierarchical organization of its parts. Secondarily, we are interested in the flow of words that support the argumentative drive.

We will combine the following steps:

- Fragmentation of the text continuum (the sequence of occurrences that comprise it) into roughly equal-length fragments while respecting the sentences. This initial fragmentation allows us to begin identifying the text's structure.
- Visualization of the trajectory of these fragments using CA. This trajectory represents the shape of the text.
- Identification of the text's hierarchical structure to follow the flow of arguments. This requires time-constrained clustering.
- Tracking the flow of arguments along the hierarchical tree, which involves searching for the *hierarchical characteristic words*, as defined in Section 8.3.4.

These steps are described in detail below.

8.3.1 Text continuum fragmentation

Often, the text to be studied is not divided into fragments. This is the case with the example used in this chapter, an oral closing argument. Therefore, we must divide it into fragments that are long enough to smooth out the high degree of local variability and reveal the text's shape. These fragments should contain at least 500 occurrences. This fragmentation can be arbitrary and correspond to an artificial division of the text into 500-occurrence fragments. Alternatively, if the text is divided into sentences (or pseudo-phrases), the fragmentation can be applied to 500-occurrence fragments while maintaining the integrity of the sentences. This is the approach taken here. These fragments will be referred to as *blocks* henceforth.

Remark. An alternative to this fragmentation method is to use a tool that can detect real breakpoints, as Bécue-Bertaut et al. (2014a) did. This method has the advantage of respecting the actual breakpoints. However, it often results in some chunks being very small yet overly influential in the CA.

8.3.2 Visualizing the evolution of the text

Based on our experience working with argumentative texts, we have found that changes in their vocabulary resemble changes in textual time series. Such a series could consist of editorials from the same newspaper or inaugural speeches from successive prime ministers of the same country, for example. These corpora typically demonstrate consistent vocabulary renewal, resulting in an arch-shaped trajectory of their components on the primary plane of CA. This pattern is commonly referred to as the *Guttman effect*, as discussed in previous chapters. In a long text, the arch pattern indicates gradual, smooth changes in vocabulary and implies well-connected blocks. Of course, we also expect departures from this basic pattern that indicate a special event altered the renewal of vocabulary. Singular blocks, flashbacks, and loops can occur, providing valuable information. The trajectory of the blocks, or the *shape of the text*, illustrates narrative and argumentative development.

8.3.3 Hierarchical structure of the text

Clustering aggregates blocks based on their coordinates along the selected CA principal axes. Time-constrained clustering ensures that only consecutive blocks or nodes are aggregated (see Chapter 4). The speaker's strategy is reflected in the hierarchical structure revealed by clustering. Thus, considering higher dimensionality with respect to words improves CA visualization.

8.3.4 Argumentative flow

The hierarchical tree illustrates the organization of the text. We characterize it by identifying words whose distribution differs significantly from that of a random sample. Specifically, we look for words that appear more frequently in a block than would be expected by chance. These words are called *characteristic words*. The test to detect them was discussed in Chapter 1. We also look for *hierarchical characteristic words* of nodes in the hierarchy, as defined below. Note that a node corresponds to a part of the text. Hierarchical words enable a lexical description of a time-constrained hierarchical tree and are computed as follows: First, we compute the characteristic words for each node of the hierarchy, including the terminal words corresponding to the blocks. Then, each word is assigned to the node that best characterizes it (i.e., the node with the smallest `p-value`). These words are displayed on the dendrogram of the hierarchical tree.

8.4 Results from the example

We use the data presented in Section 2. The results were obtained using the script **chap8**.

8.4.1 Text preparation

The oral closing argument was transcribed with correct spelling but with syntactic errors and missing punctuation in certain passages. A classical reading of the transcription made it possible to divide the speech into 191 short *equivalent-to-sentences*. These sentences were then grouped into 19 blocks of similar length. Consequently, the shortest block contains 463 occurrences, while the longest contains 631. The corpus contains a total of 10,403 occurrences (not including numbers). The corpus consists of 1,800 different words.

8.4.2 Vocabulary

Glossaries of words and repeated segments of consecutive words provide an initial quantitative overview of the corpus's content. Examining words and repeated segments that have been used at least five times can reveal peculiarities of the case. As usual, the most frequent words are functional words. The most frequent full word is *insurance*, appearing 51 times. The prosecutor confidently declares *We have no incriminating evidence, but that doesn't pose much of a problem*, yet he demonstrates that many facts and data, when logically combined, constitute clues. He also says *We are not talking about suspicions. I don't want to give the defense this opportunity. We are talking about clues. We are talking about perfectly contrastable clues.* According to the cited case law, circumstantial evidence is sufficient for a conviction. The prosecutor leaves no room for doubt, delivering his speech with a firm, confident tone.

Note that some words appear regularly throughout the closing argument, while others appear in close succession, indicating localized use. Examples of the former include words such as *perfectly, exactly, totally, obviously* and *we know*. These words give the prosecution an assured, demonstrative tone. Their frequent use creates the impression that the defendants are guilty. However, there is no material evidence, and the only direct witness who overheard the defendants discussing the crime has recanted. Conversely, words such as *witnesses, testimonies, statements, manifest* and *facts* appear regularly to demonstrate that the prosecutor's closing argument is based on solid evidence and proven facts.

Conversely, some words appear in only a few parts of the corpus. Notably, the words *witness* and *statement* are used much more locally than *witnesses* and *statements*. What can we deduce from this? First, lemmatization before editing would not have revealed this phenomenon. Further analysis and a return to the text are necessary to understand the meaning of this information. Among the witnesses, we find the relevant one—the woman who overheard the conversation between the defendants. Among the statements, we find hers.

8.4.3 Shape of the closing argument unveiled by CA

A total of 302 words that appeared at least five times were selected for the application of CA to the aggregate blocks × words table. The total inertia of this table is 1.2. Seven axes have above-average inertia, with respective eigenvalues of 0.129, 0.125, 0.093, 0.085, 0.081, 0.079, and 0.071. Together, these axes account for 54.8% of the total inertia.

When dealing with a highly structured argumentative text, one would expect to see well-differentiated yet connected parts. This is because the arguments must be linked, resulting in a gradual renewal of vocabulary. In this case, rearranging the columns of the document-word matrix creates a table with a charged diagonal. CA maps represent this structure by arranging blocks along an approximate parabola or arch pattern. The deviations from the parabola contain the information, and the parabola serves as a reference model.

Now, let's take a look at what happens in the example. Figure 8.1 (top) plots the trajectory of the blocks on the main CA plane. Despite the fact that this initial plane retains only 21% of the total inertia, the visual representation effectively captures most of the temporal structure. This is evident from the absence of the Guttman effect on any axis higher than the second. The first two axes each retain a similar percentage of inertia (10.68% and 10.32%, respectively). Therefore, the graph should be interpreted without favoring any particular direction. These percentages are much higher than the 7.70% inertia retained by the third axis. This emphasizes the importance of the trajectory shown on the first CA plane.

First, we will examine the closing argument's structural organization, as reflected in its trajectory and hierarchical composition. At this stage, we will focus only on the organization of the text, not its content, which is defined by the words used to express it. The first map (see Figure 8.1, top) illustrates the trajectory of well-represented blocks in terms of CA representation quality with a solid black line. The line connecting poorly represented blocks is gray and dotted. The trajectory from BL1 to BL9 is highly regular, indicating that the text is delivered steadily and in a controlled manner. However, progress slows down completely from BL9 to BL12. These last blocks' lexical peculiarities are poorly reflected on the first CA plane, in accordance with their poor representation quality, as calculated by the CA interpretation aids. BL13 marks an important leap, suggesting that it introduces a new argument that resumes the progression of the closing argument. This is evident in the trajectory of BL13 through BL16 (excluding BL14), which extends the previous parabola (see Figure 8.1, top).

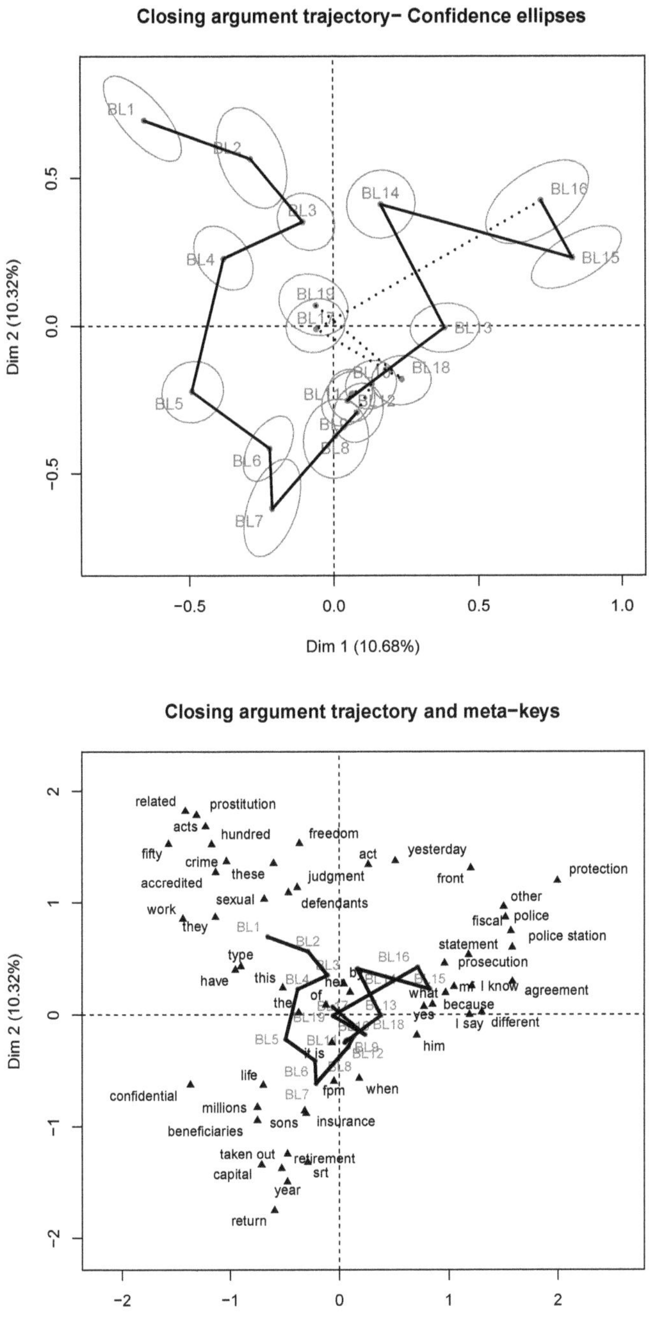

Figure 8.1: First CA plane. Top: Documents enriched by confidence ellipses; bottom: Most contributive words.

The words that contribute the most to the first CA factorial plane are shown at the bottom of Figure 8.1. The information they provide is discussed below, along with the clustering results. Figure 8.2 shows some of the repeated segments on the same plane.

Below, we will review the information from the CA's planes and the results of the clustering analysis (see Section 8.4.4). Based on this information, we must determine what is happening between blocks BL10 and BL12, why there is a significant advance at block BL13, and why the progression is disordered between blocks BL17 and BL19. These peculiarities in the progression pattern suggest difficulty in delivering the closing argument in an orderly manner from block BL9 onwards.

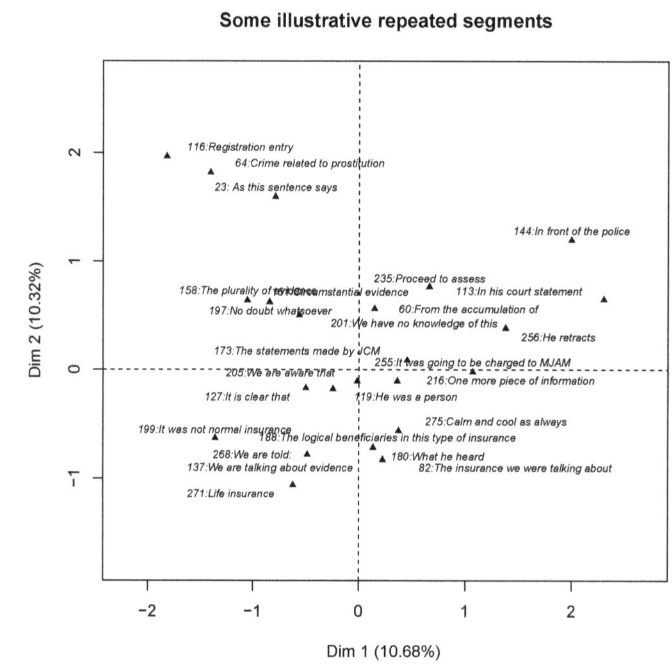

Figure 8.2: First CA plane. Repeated segments as illustrative elements.

8.4.4 Time-constrained hierarchical clustering

The relationship between words and blocks is not fully captured by the first CA factorial plane. To account for the higher dimensionality, time-constrained hierarchical clustering is applied using the coordinates of the blocks on the first three axes. According to the RV coefficient (Josse et al., 2008), the subspace defined by the first three CA axes best captures the chronological evolution of the blocks. The RV coefficient measures the similarity between the configuration of the blocks in any CA subspace and the configuration induced by chronology. The latter variable is the rank of the first record of each block. This coefficient takes the highest value (.39) when the subspace supported by the first three axes is considered. Together, the first three axes retain 28.7% of the inertia, indicating that factors other than chronology influence vocabulary choice.

With the hierarchical tree (see Figure 8.3), we are not looking for a partition of the closing argument. Instead, we are interested in its overall organization. This tree is labeled to illustrate the progression of the arguments, which will be discussed in the next section.

8.4.5 Evolution of the argument over time

We describe the time-constrained hierarchical tree using the hierarchical words, which are selected as explained in Section 8.3.4. This labeled hierarchy enables us to trace the arguments represented by the words. For example, the Table 8.1 lists the hierarchical words corresponding to the node [BL5 BL6 BL7 BL8].

Table 8.1: Extract of the list of hierarchical words associated with the node containing [BL5] to [BL8].

```
> round(res.Label$"BL5 BL6 BL7 BL8",2)
              Intern % glob % Intern freq Glob freq  p.value v.test
life              1.09   0.32          19        26     0.00   5.41
sons              0.86   0.26          15        21     0.00   4.67
case              0.69   0.21          12        17     0.00   4.08
millions          0.57   0.16          10        13     0.00   3.98
beneficiaries     0.52   0.14           9        11     0.00   3.96
the               2.23   1.41          39       113     0.00   3.05
beating           0.29   0.07           5         6     0.00   2.82
recorded          0.80   0.39          14        31     0.01   2.74
are               0.63   0.27          11        22     0.01   2.73
all               0.40   0.14           7        11     0.01   2.72
her               1.09   0.67          19        54     0.03   2.14
```

The word *evident* aptly summarizes the first sequence, which presents objective and indisputable facts (see Figure 8.3). These include details of the life insurance policy, such as the amount and beneficiaries, as well as other information, such as the fact that the main defendant is married to another woman. The second sequence uses circumstantial evidence to prove the defendants' guilt in the murder of MJAM, and is therefore more speculative. According to case law, a guilty verdict can be obtained in the absence of direct evidence of the crime if there are a number of concurrent and conclusive clues of the crime. This is the case here. Therefore, the prosecutor's reasoning is well summed up by *I say* (see Figure 8.3).

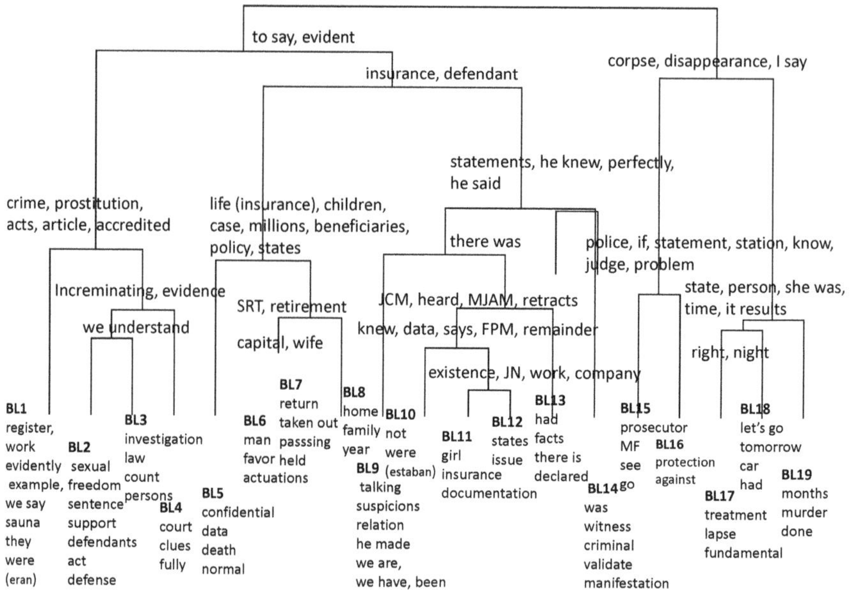

Figure 8.3: Labeled hierarchical tree with contiguity constraints.

To make reading easier, view the first CA factorial plane and the labeled tree simultaneously. Follow the divisions corresponding to the three main sequences: [B1-BL4], [BL5-BL14], and [BL15-BL19]. The first sequence is in the second quadrant of the first CA factorial plane. The second sequence primarily occupies the third and fourth quadrants of the same plane. The third sequence has a disordered trajectory in the first quadrant and the central part.

BL1-BL4 : The crimes with which the two accomplices will be charged—prostitution, murder, and insurance fraud—are specified at the beginning. The latter two are closely related. The prostitution charge is clearly substantiated by the *sauna*'s register as discussed at the beginning of BL1 and BL2. Regarding the other offenses, the prosecutor claims that despite the lack of *incriminating evidence*, he has sufficient *clues* that are *rationally related to the facts to be proven* to *fully* convince the *court*. The prosecutor presents the legal framework and the sections of the law on which he will rely in his closing argument. *Evidence* and *clues* characterize BL3 and BL4, respectively.

BL5-BL14 : This sequence is central to the closing argument and will determine the outcome of the case. First, [BL5-BL8] outlines the undisputed fact that the insurance had *a very high payout*, was *not at all usual policy*. The beneficiaries are the defendant's children, not the victim's. The prosecutor emphasizes the implausibility of the sentimental relationship between the main defendant and the victim. Then, the prosecutor moves on to the more challenging section, [BL10-BL13]. He fears that he will not be able to present the incriminating witness's police statement convincingly. The woman said she had *heard* the accomplices discussing the next murder, but she retracted her statement before the judge (last sentence of BL12 and BL13). The prosecutor is slow to mention this, and the case does not progress. He spends a lot of time discussing what the defendant knows, what he should know, and whether MJAM was his real companion. However, he does not provide any new information. Finally, at the end of BL12, he abruptly announces *We are going to review MF's statement* and continues with this topic throughout BL13. His difficulty is reflected in rambling sentences and word repetitions. For example, he repeats the word *heard* six times in only two consecutive sentences in [BL13]. Then, [BL14] changes direction. The prosecutor asks the court to validate the witness statement, using the words *witness,manifestation* and *validate.* Once again, he uses words from BL3, where the prosecutor announced that he would use a circumstantial argument. In short, there is a lack of advancement in [BL10-BL12], followed by a leap in BL13 and an inversion of the trajectory in BL14.

BL15-BL19 : this sequence is very disorganized. First, [BL15-BL16] focuses on witness MF, her statement to the police, and her subsequent retraction. A new vocabulary emerges. Words such as *police, testimony, judge* appear frequently in these two blocks. New information has come to light, including the time span between the body being found and the insurance payout being claimed. It has also emerged that the defendant was not involved in discovering the victim's death. [BL17-BL19] shifts abruptly toward [BL10-BL12] and concludes close to the CoG (see Figure 8.1). This corresponds to the use of words belonging to earlier topics. In fact, the prosecutor revisits points already covered to provide new information. *Another piece of data* is a refrain in this section. The prosecutor again mentions the victim, specifying that she was mentally ill—a new argument in favor of the implausibility of her relationship with the sauna owner, FPM. The prosecutor mentions that the victim was socially isolated and undergoing psychiatric *treatment*, a fact of which her so-called romantic partner was unaware. The haphazard nature of the closing argument seems to indicate the prosecutor's exhaustion and lack of control. Finally, he briefly concludes with the formal charges in [BL19].

It should be noted that the prosecutor is young and inexperienced, and this is his first challenging case. Even before the trial began, his colleagues told him he would lose. Determined to win, he knew that if he failed to convince the judges, the defendants would not only go free, but also collect life insurance.

In conclusion, the judge agreed with the prosecutor's recommendation and gave the defendants lengthy prison sentences. This decision was upheld by the Supreme Court upon appeal. The prosecutor was at a disadvantage due to the lack of evidence, the victim's status as a prostitute, and the financial resources of the killer, who hired one of the best criminal defense lawyers.

The outcome of the trial confirmed the effectiveness of the prosecutor's strategy. Despite some disorganization in the final section, the prosecutor presented his closing argument in a well-structured and progressive manner. He demonstrated good control of his lexical choices by consistently using words that conveyed conviction.

8.5 Conclusion

This chapter does not introduce new methods. Instead, it presents an original application of well-known text analysis methods. By combining CA, clustering methods, and characteristic word extraction, one can understand the argumentative discourse embedded in a text's flow and vocabulary choices. CA visualizes a text's structure and reveals its organization through clustering. The argumentative flow is traced through this hierarchical structure, revealing its peculiarities. In our example, the methodology reveals the organization of the closing argument. First, the arguments undergo a controlled evolution. Then, the evolution becomes irregular, reflecting the prosecutor's increasing difficulty in organizing his argument. This process can be applied to both oral and written texts. CA produces especially good results when studying heterogeneous argumentative texts. In such cases, each part plays a specific role, primarily defined by its differences from the others. This statistical approach can benefit other areas where persuasion, whether oral or written, is relevant to decision-making.

9

Extension of multivariate statistical methods to multilingual corpus

This chapter is the work of Belchin Kostov, Ramón Alvarez-Esteban, Mónica Bécue-Bertaut and François Husson

9.1 Introduction

This chapter examines a method for analyzing open-ended survey questions answered in different languages. The starting point is as follows: All respondents completed the same closed questionnaire and answered a common open-ended question in their respective languages. The goal is to analyze all of the open-ended responses together by examining the relationship between word choice and the closed-ended answers. In Chapter 7, we discussed how to analyze open and closed questions together by giving the latter an explanatory role. This process involves creating a GALT which summarizes the associations between words and variables. When working with multiple samples, a GALT must be created for each sample. Then, they are all analyzed simultaneously as a global table using MFA, which is tailored to process multiple GALTs. The result is *multiple factor analysis for generalized aggregate lexical tables* (MFA-GALT).

The chapter is organized as follows: Section 9.2 presents a motivating example extracted from a real survey. Section 9.3 presents the data structure and the notation. Section 9.4 is devoted to MFA adapted to multiple GALTs (MFA-GALT) and Section 9.5 exposes the properties of the method. Eventually, the application of MFA-GALT to the example (Section 9.6) shows its capabilities. The main conclusions are presented in Section 9.7.

DOI: 10.1201/9781003397472-9

9.2 Example: multilingual survey

A railway company conducted a survey to determine passenger satisfaction with its night trains. Passengers were asked to rate their satisfaction with 13 aspects related to comfort. These aspects included general comfort, cabin comfort, bed comfort, seat comfort, and the cleanliness of common areas, cabins, and toilets. Passengers were also asked to rate the attentiveness of the staff, including their initial welcome and the attention they provided during the trip. Other aspects included language proficiency, cabin spaciousness, and air conditioning. Each aspect was scored on an 11-point Likert scale from 0 (very poor) to 10 (excellent). Additionally, passengers were asked to respond to an open-ended question about areas in need of improvement. They were asked to respond spontaneously in either English or Spanish.

9.3 Data structure and notation

L samples have answered a questionnaire with closed questions, either quantitative or categorical, all of identical type, which constitute the contextual data. These samples have also answered an open-ended question in their respective languages. These answers correspond to the textual data. Sample l has I_l respondents who use J_l different words in language l. From these responses, we constructed Table $\mathbf{Y_l}$, an $I_l \times J_l$ respondents $\times$ words matrix. N_l is the grand total for this table. From $\mathbf{Y_l}$, we compute the $(I_l \times J_l)$ proportion table $\mathbf{P_l} = \mathbf{Y_l}/N_l$.

The answers to the closed-ended questions shared by all samples are encoded in the $(I_l \times K)$ $\mathbf{X_l}$ table. The columns of $\mathbf{X_l}$ correspond to quantitative or dummy variables. The latter encode the categories of one or more categorical variables. Regardless of the type, k and K denote the column variable k and the total number of column variables, respectively. In what follows, the term *variable* will be used for both types. Figure 9.1 summarizes the data stucture.

If we consider only sample l, the weights of the respondents are obtained from the column margin of $\mathbf{P_l}$. They are proportional to the length of their free responses. These weights are stored in the $(I_l \times I_l)$ diagonal matrix $\mathbf{D_{I_l}}$. The total weight of the respondents in the same sample equals one. Similarly, the weights of the words are obtained from the row margin of $\mathbf{P_l}$ and are thus proportional to their counts. These weights are stored in the $(J_l \times J_l)$ diagonal matrix $\mathbf{D_{J_l}}$. The total weight of the words used by the same sample also equals one. $\mathbf{X_l}$ is centered and possibly reduced to a unity variance in the case of quantitative variables, for the weighting system $\mathbf{D_{I_l}}$.

The data structure, that captures the relationships between words and variables, is the $(J_l \times K)$ GALT calculated as $\mathbf{Y_{A_l}} = \mathbf{Y_l^T X_l}$. From this matrix, we calculate

$$\mathbf{P_{A_l}} = \frac{\mathbf{Y_{A_l}}}{\mathbf{N_l}} = \mathbf{P_l^T X_l}.$$

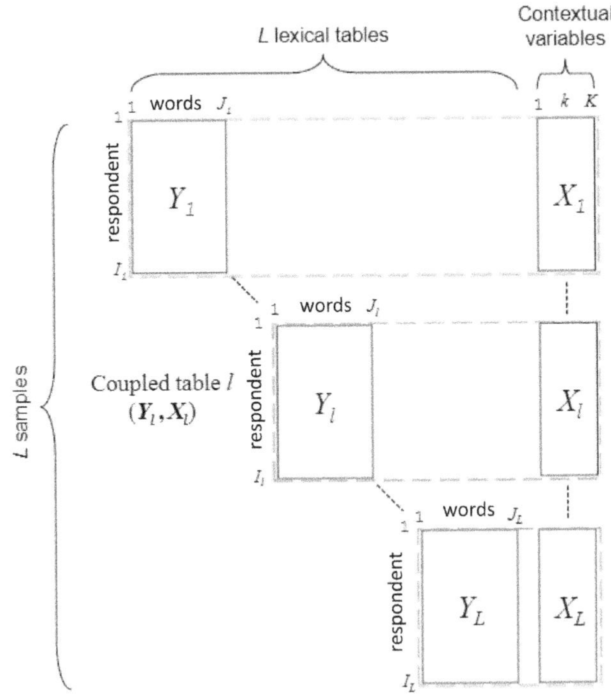

Figure 9.1: Sequence of L coupled tables.

In the global analysis of the L samples, we have to deal with $I = \sum_l I_l$ respondents who have used $J = \sum_l J_l$ different words in the $N = \sum_l N_l$ occurrences that they pronounced through all the free answers. The respondent and word weights are resized so that both totals are equal to 1 for, respectively, the I respondents and J words. To this end, the respondent and word weights are multiplied by N_l/N in sample l. The respondent global weights are stored in the $(I \times I)$ diagonal matrix $\mathbf{D_I}$. The word global weights are stored in the $(J \times J)$ diagonal matrix $\mathbf{D_J}$.

The $(I \times K)$ global table $\mathbf{X}$ is obtained by juxtaposing the L tables $\mathbf{X_l}$ column-wise, centered by set. Consequently, table $\mathbf{X}$ is also centered for the $\mathbf{D_I}$ weighting system.

We assume that $K < J$. The symbols I, I_l, J, J_l, K, and L refer to both the set and its cardinal number.

9.4 MFA on multiple GALTs

The methods combined in MFA-GALT, MFA and CA-GALT were presented in Chapters 6 and 7, respectively. The L separate tables are of the GALT type and are created from L coupled tables $(\mathbf{Y_l}, \mathbf{X_l})$, $l = 1, \dots, L$. This raises the question of how to adopt an MFA approach for analyzing multiple GALTs.

MFA is typically applied to a set of rows described by multiple sets of columns. However, we now need to analyze several sets of row words described by one set of column variables. Fortunately, we are in a CA-type context, so interchanging the roles of the rows and columns does not affect the outcome.

Similar to classical MFA, MFA-GALT is performed in two steps. First, each subtable —in this case, a GALT— is analyzed separately using the corresponding factorial method —in this case, CA-GALT. In the second step, a global factorial analysis is performed on the multiple GALTs. This addresses each set as it was dealt with in its separate analysis, but it uses reweighting to balance the influence of the sets. This reweighting involves dividing the weights of the rows of set l by the first eigenvalue obtained in the separate analysis of that set. The two steps are detailed below.

First step: separate analyses

Separate CA-GALT calculations are performed on each GALT, according to the scheme in Section 7.5. The only difference is the metric used in the row space. In this case, in all separate analyses, the covariance/correlation matrix calculatd from all the respondents $\mathbf{C} = (\mathbf{X^T D_I X})$ is used instead of the matrices $\mathbf{C_l} = (\mathbf{X_l^T D_{I_l} X_l})$. This is due to the need to place all the row sets in the same metric space. Accordingly, $\mathbf{C^{-1}}$ is used to standardize $\mathbf{Y_{A_l}}$. $\mathbf{C^{-1}}$ is substituted by the Moore-Penrose pseudoinverse $\mathbf{C}^-$ if $\mathbf{C}$ is not invertible. In this first step $\mathcal{T}_l = \mathbf{D_{J_l}^{-1} P_{A_l} C^-}$ is analyzed via PCA($\mathcal{T}_l$, $\mathbf{C}$, $\mathbf{D_{J_l}}$). The first L eigenvalues, λ_1^l, calculated in separate analyses, will be used in the second step. We recall that the eigenvalue is equal to the square of the singular value. Similarly, the calculations could be performed using GSVD($\mathcal{T}_l$, $\mathbf{C}$, $\mathbf{D_{J_l}}$). This is merely a question of terminology. As we saw in Chapter 2, the two calculation methods produce equivalent results. However, eigendecomposition is the more standard method in the field of MFA.

Second step: global analysis

The row weighting system is updated to balance the influence of each set within the global analysis. Matrix $\mathbf{D_J}$ is divided into L blocks, by construction. Block l corresponds to the J_l words used in sample l. The weights of the words of block l are divided by λ_1^l, the first eigenvalue of the separate analysis of subtable l. The resulting weights are stored into the $(J \times J)$ matrix $\mathbf{D_{J_\lambda}}$.

The $(J \times K)$ multiple table GALT $\mathbf{P_A}$ juxtaposes the L matrices $\mathbf{P_{A_l}}$ column-wise but resized by multiplying them by the coefficient N_l/N, i.e., $(\mathbf{P_{A_l}} \times N_l/N)$. Applying double standardization to the $\mathbf{P_{A_l}}$ table, by both the rows and columns, results in the $(J \times K)$ table. $\mathcal{T} = \mathbf{D_{J_\lambda}^{-1} P_A C^{-1}}$. If $\mathbf{C}$ is not invertible, $\mathbf{C^{-1}}$ is substituted by $\mathbf{C^{-}}$. Then, MFA-GALT is performed through PCA($\mathcal{T}$,$\mathbf{C}$,$\mathbf{D_{J_\lambda}}$). A thorough explanation of the methodology can be found in Kostov et al. (2024).

9.5 Main properties of MFA-GALT

9.5.1 Representation of the row-words and the column-variables

MFA-GALT provides the classical outputs of the factorial methods:

- The eigenvalues λ_s are stored in the $(S \times S)$ diagonal matrix Λ.
- The dispersion axes in the row space, v_s, are stored in the columns of the $(K \times S)$ matrix $\mathbf{V}$. The S factors on the rows are stored in the columns of the $(J \times S)$ matrix $\mathbf{F}$, which is calculated as $\mathbf{F}=\mathcal{T}\mathbf{CV}$.
- The S factors on the columns are stored in the columns of the $(K \times S)$ matrix $\mathbf{G}$. They are calculated based on the transition relationships between the factors on the rows and the factors on the columns. $\mathbf{G} = \mathcal{T}^{\mathbf{T}}\mathbf{D_{J_\lambda}F}\Lambda^{-1/2}$.

9.5.2 Superimposed representation of the global and partial clouds of variables

According to the L sets of row words, the column variable k of $\mathcal{T}$ can be divided into L subcolumns, which are named *partial* variables and denoted k^l, $k = 1, \ldots, K$. It is interesting to simultaneously represent the L partial scatter plots, each made up of the corresponding K partial variables, as well as the global one on the same referential axes. To this end, we consider the L matrices $\mathcal{T}_l$, with dimension $(J_l \times K)$, derived from matrix $\mathcal{T}$ by retaining only the row words belonging to set l. From these matrices, the $(J \times K)$ matrices $\tilde{\mathcal{T}}_l$ are built by completing $\mathcal{T}_l$ with 0 to be the dimension of $\mathcal{T}$.

To be represented on the global axes, the K partial variables corresponding to the set l are considered as supplementary columns in the global analysis. Their coordinates are calculated using the transition relationships and stored in the $(K \times S)$ matrix $\mathbf{G^l}$:

$$\mathbf{G^l} = \tilde{\mathcal{T}}_l{}^{\mathbf{T}}\mathbf{D_{J_\lambda}F}\Lambda^{-1/2}.$$

Thus, the coordinates of the partial variable k^l corresponding to set l are calculated from the coordinates of the words used only by sample l. This is expressed very simply thanks to the structure of matrix $\tilde{\mathcal{T}}_l$ which contains only 0 except for the rows belonging to set l:

$$G_s(k^l) = \frac{1}{\sqrt{\lambda_s}} \frac{1}{\sqrt{\lambda_1^l}} \sum_{j \in J_l} \tau_{jk} d_{J_{jj}} F_s(j).$$

We recognize the usual transition formula here, but it is restricted to the words in group l. In this formula, the term τ_{jk} denotes the generic term of $\mathcal{T}$ and $\frac{1}{\sqrt{\lambda_1^l}} d_{J_{jj}}$ denotes the generic term of matrix $\mathbf{D}_{\mathbf{J}_\lambda}$, being $d_{J_{jj}}$ the initial weight of the word j (see Section 9.3). According to the last equation, the partial variables relative to set l are on the side of the words that respondents with high values for these variables overuse. When the vocabularies used by the different samples produce similar structures for the variables, the partial homologous points k^l are close to each other. The variables whose homologous points are distant constitute an exception to this common structure.

Typically, all the *partial* variables can be represented in a single scatter plot. This illustrates the similarities and differences in the structure of the variables among the samples. Global variables can also be added to this representation.

9.5.3 Synthetic representation of the sets on the same graph

Another result is the representation of the L sets of words in the same graphs. In these graphs, each set is represented by a single point. These graphs' axes are interpreted similarly to the global axes. Without delving into the calculation details, we can say that the sets are closer together in this graph the more similar their variable structures are to one another. Specifically, the value of set l coordinate on axis s is equal to the sum of the contributions of words in set l to this axis.

A small distance between two sets on an axis indicates that the sets include the structure expressed by the axis with the same intensity. This intensity is low when the sets' coordinates are small and high when they are large. The representation of the sets shows how similar or different they are in terms of global analysis factors.

9.5.4 Association between vocabulary and contextual variables

Measuring the association between vocabulary and contextual variables is useful for selecting meaningful words and interpreting results. The successive measures performed on each sample are detailed in Bécue-Bertaut and Pagès (2015).

Individuals	Score variable	Words		
		word A	word B	word C
ind 1	4	1	0	1
ind 2	6	1	0	0
ind 3	3	0	1	1

Individuals	Words	Score variable
ind 1	word A	4
ind 1	word C	6
ind 2	word A	4
ind 3	word B	6
ind 3	word C	3

Figure 9.2: Reorganization of the data for the one-way ANOVA that measures the association between vocabulary and a contextual variable.

In brief, vocabulary is said to be associated with a variable if the words differ significantly from each other in the values taken by the individuals using them. The association between a categorical variable and vocabulary is evaluated using a chi-square test on the frequency table that crosses words and categories (i.e., an ALT). For quantitative variables, a one-way ANOVA is used. Before computing this ANOVA, however, the data table is reorganized as shown in Figure 9.2. Each row corresponds to an occurrence, i.e., a word cited by one individual. Then, the score and word variables have as many values as occurrences. Finally, a one-way ANOVA is performed to detect relations between vocabulary and scores.

Note that since these occurrences are not independent, the assumptions of ANOVA do not apply. Therefore, a permutation test is preferable.

9.6 Results from the example

We use the data presented in Section 2. The data is stored in the data structure shown in Figure 9.3. The results were obtained using the script **chap9**.

Data preprocessing includes careful spelling correction. Stopwords are removed and then the words used at least 10 times are kept for the Spanish corpus (=all answers given in Spanish) whereas the threshold is 5 for the English corpus (Lebart et al., 1998; Murtagh, 2005). Of the respondents, 977 from the Spanish sample and 283 from the English sample provided an answer. The average response length is 3.1 in both cases. The Spanish corpus contains 3,029 occurrences of 88 distinct words, while the English corpus contains 871 occurrences of 68 distinct words.

Missing values were imputed for the score variables. The grading scale was inverted so that the highest values correspond to the greatest dissatisfaction, making the graphs easier to read.

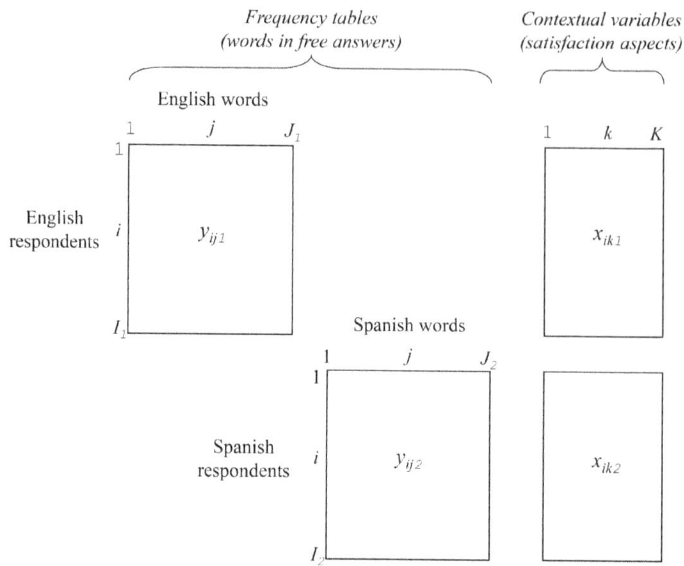

Figure 9.3: The dataset. On the left, the lexical tables; on the right the contextual variables tables. In the example, $I_1 = 283$ (English respondents), $I_2 = 977$ (Spanish respondents), $J_1 = 68$ (English words), $J_2 = 88$ (Spanish words), $K = 13$ (satisfaction scores).

9.6.1 Preliminary findings

The most frequent words provide an initial overview of the complaints, which are comparable in both languages and expressed with similar vocabulary. The *space* is too small, and there is no place for *luggage*. The *cabins* and *seats* are *uncomfortable*, and the *toilets* need *to be cleaned more often*. The *air conditioning* seems problematic. In the English sample, the words *staff* and *English* are frequently mentioned. Aspects that were not requested to be evaluated are cited, such as *price*.

Table 9.1 provides an initial overview of the means and standard deviations of the satisfaction scores. *Staff trip attention* obtains the highest score (8.07) for Spanish speakers, and English speakers gave the highest score to *Cabin cleanliness* (7.59). The lowest score was *Cabin room* for both Spanish (5.33) and English speakers (5.71). The three aspects related to staff (*staff welcome, staff attention during the trip, and staff language skills*) were clearly less appreciated by English speakers than by Spanish speakers.

The association between vocabulary and a contextual variable, columns *ass.ratio (p-value)* in Table 9.1, shows that *Air conditioning* obtains the highest ratio for both Spanish (0.107) and English (0.226) speakers. *Toilet cleanliness* ranks second for this indicator for Spanish speakers (0.090) while *Staff language skill* is placed second with 0.154 in the case of English ones, although

Table 9.1: Mean satisfaction scores and association with vocabulary ratios.

Satisfaction scores	Spanish respondents mean (SD)	Spanish respondents ass.ratio (p-val.)	English respondents mean (SD)	English respondents ass.ratio (p-val.)
General comfort	6.82 (1.80)	0.062 (0.000)	6.66 (1.95)	0.091 (0.148)
Cabin comfort	6.37 (2.07)	0.063 (0.000)	6.37 (2.03)	0.121 (0.010)
Cabin room	5.33 (2.43)	0.089 (0.000)	5.71 (2.35)	0.136 (0.000)
Bed comfort	6.70 (1.98)	0.050 (0.000)	6.72 (2.03)	0.063 (0.918)
Seat comfort	6.10 (2.20)	0.059 (0.000)	5.99 (2.38)	0.123 (0.010)
Air conditioning	6.55 (2.55)	0.107 (0.000)	6.51 (2.71)	0.226 (0.000)
Common areas cleanliness	7.41 (1.92)	0.043 (0.000)	7.54 (1.86)	0.082 (0.548)
Cabin cleanliness	7.59 (1.88)	0.056 (0.000)	7.59 (1.81)	0.116 (0.036)
Toilet cleanliness	6.21 (2.55)	0.090 (0.000)	6.29 (2.40)	0.150 (0.000)
Staff welcome attention	7.99 (1.92)	0.040 (0.018)	7.29 (2.45)	0.108 (0.062)
Staff trip attention	8.07 (1.85)	0.038 (0.048)	7.34 (2.29)	0.092 (0.294)
General aspects	7.77 (1.65)	0.038 (0.034)	7.48 (1.91)	0.079 (0.590)
Staff language skills	7.72 (2.08)	0.052 (0.000)	7.14 (2.52)	0.154 (0.000)

Toilet cleanliness, with 0.150, is third and not far behind. Note that *Staff language skill* is only ranked eighth for this indicator in the case of Spanish speakers. *Cabin room* is ranked the third for Spanish speakers and the fourth for English speakers.

9.6.2 MFA-GALT on the multilingual dataset

The ranking of aspects based on the association-with-vocabulary ratio does not coincide with the average score ranking. This implies that the aspects passengers think should be improved the most are not the same as the aspects with which they are least satisfied. Therefore, we can say that it is important to collect information through open-ended questions because they provide additional insight.

The two GALTs are built and MFA-GALT is applied. The total inertia is 9.91. The first eigenvalue, 1.75, corresponds to 17.70% of the total inertia and nearly matches the number of sets. This suggests that the two sets share a dispersion direction corresponding to the first global axis. The second (1.42, 14.36%) and third (1.23, 12.39%) eigenvalues are similar. The subsequent eigenvalues are much smaller. To avoid overemphasizing the example, we will only interpret the first two axes. For a more detailed description of the results, particularly the third dimension, see Kostov (2015).

9.6.2.1 Global representation of the satisfaction scores and words

MFA-GALT provides graphical results, with each variable pointing to the associated words. Thus, it indicates defects in the scored aspect that are or are not shared by both languages. Figure 9.4a illustrates the most well-represented satisfaction scores on the initial MFA-GALT principal plane through their covariances with the axes. To avoid overloading the graphics, only the most well-represented scores are shown (here, those with a summed square cosine of at least 0.5 on the two axes). First, we examine the global representations

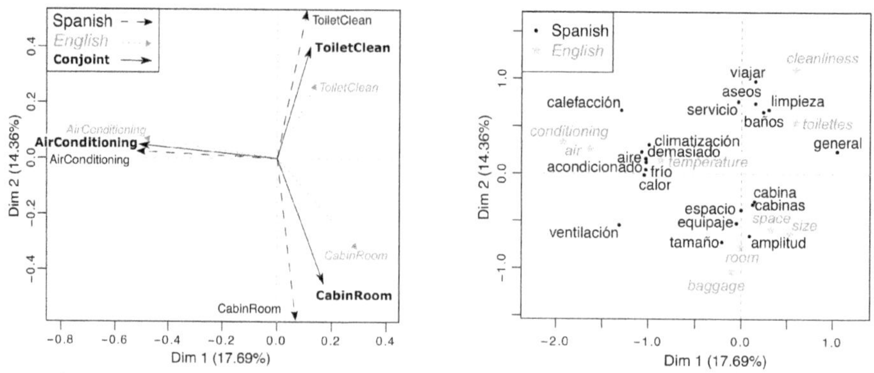

(a) Scores on MFA-GALT plane

(b) Words on MFA-GALT plane

Figure 9.4: MFA-GALT Representation of scores (left) and words (right) on the plane (1,2).

of the scores, which reveal a three-polar structure. These three poles refer to inconveniences associated with *Air conditioning*, lack of *Toilet cleanliness* and problems related to *Cabin room*.

This aligns well with the association-with-vocabulary ratio suggestions. Figure 9.4b shows Spanish and English words that contribute more than twice the average. The words highly associated with *air conditioning*, are shown, revealing its drawbacks: *Air/aire*, *conditioning/acondicionado*, *temperature*, *frío* (=cold) *climatización* (=air conditioner), *ventilación/ventilation* and *calefacción* (=heating). On the positive side of the second axis, *cleanliness/limpieza*, *toilettes/aseos/baños* characterize the lack of toilet cleanliness. On the negative side, problems with *Cabin room* are described with the words *size/espacio* and *cabins/cabina/cabinas.*

In this example, axis 3 is specific to one set: the English set, which has problems with staff who speak poor English.

9.6.2.2 Partial representation of the satisfaction scores

Figure 9.4a shows the superimposed global and partial representations of satisfaction scores on the (1,2) plane. It illustrates the similarities and differences between the two sets with regard to the association between words and scores. On the first axis, *Air conditioning* exhibits similar behavior in both sets. On the second axis, *Toilet cleanliness* and *Cabin room* are more strongly associated with Spanish vocabulary than with English, resulting in higher covariances. Complaints about the former seem more accentuated, resulting in more words. Only English speakers complain about the lack of *staff language skills* on the third dimension.

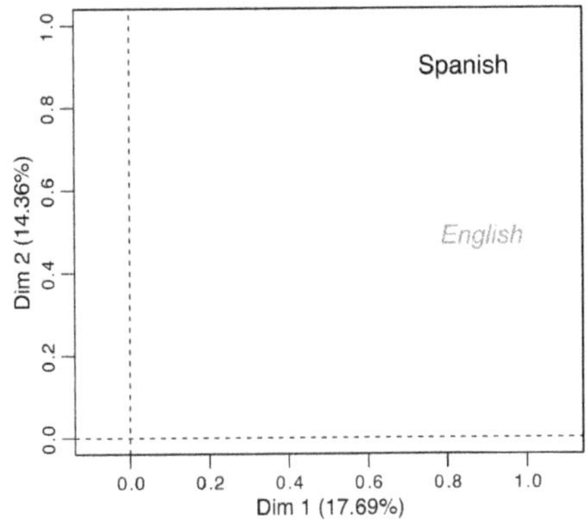

Figure 9.5: Representation of the sets.

9.6.2.3 Representation of the sets

Similarity measures confirm that both sets share dispersion directions. The RV coefficient, a multivariate generalization of the squared Pearson correlation coefficient, is 0.74 ($p < 0.001$). This result suggests that the partial configurations are close yet not homothetic.

According to their representation on the first dimension, the coordinate of the Spanish sample is 0.85. The coordinate of the English sample is slightly higher, at 0.91 (see Figure 9.5). This indicates that the first axis provided by MFA-GALT is highly important for both sets. Consequently, this is a common dispersion structure. However, the Spanish set has a much larger coordinate on the second axis (0.91 vs. 0.51). This suggests that the second MFA-GALT axis is more significant for Spanish speakers than for English speakers. The opposite is observed for the third axis (0.42 for Spanish speakers versus 0.81 for English speakers).

9.7 Conclusion

This chapter describes a principal component method for answering open-ended questions across different languages. This type of textual and contextual data yields a series of coupled tables. Each of these tables consists of a frequency table, also known as a lexical table, and a quantitative/qualitative table. We analyze these data by examining the relationships between words and contextual variables. We combine two methods, CA-GALT and MFA, hence the name of the method: *Multiple Factor Analysis for Generalized Aggregated Lexical Tables* (MFA-GALT). The first method places the words of the different sets in the same space generated by the variables, resulting in the construction of GALTs. The second method allows us to analyze these tables simultaneously while maintaining MFA properties.

Using MFA-GALT, we analyzed an international survey containing open-ended questions answered in two different languages. This allowed us to study the following:

- Similarities among words from the same language.
- Similarities among homologous words from different languages.
- Associations between words and satisfaction scores.
- Similarities between satisfaction score structures (partial representations).

The results demonstrate that MFA-GALT effectively synthesizes data into easily interpretable outputs.

Bibliography

Alvarez-Esteban, R., Bécue-Bertaut, M., and Valencia, O. (2004). Étude de la stabilité des valeurs propres de l'afc d'un tableau lexical au moyen de procédures de reéchantillonage. In *Proceedings of the 7th International Conference on the Statistical Analysis of Textual Data (JADT-2004)*, 42–51. JADT-2004, Louvain-la-Neuve, Belgium.

Alvarez-Esteban, R., Valencia, O., and Bécue-Bertaut, M. (2010). Assessing the stability of supplementary elements on principal axes maps through bootstrap resampling. contribution to interpretation in textual analysis. In *Advances in Data Analysis*, 3–11. Birkhäuser Boston, Boston. doi:10.1007/978-0-8176-4799-5_1.

Arbelaitz, O., Gurrutxaga, I., Muguerza, J., Pérez, J.M., and Perona, I. (2013). An extensive comparative study of cluster validity indices. *Pattern Recognition*, 46(1): 243–256. doi:https://doi.org/10.1016/j.patcog.2012.07.021.

Basu, S., Davidson, I., and Wagstaff, K., eds. (2008). *Constrained Clustering*. Chapman & Hall/CRC Data Mining and Knowledge Discovery Series. Chapman & Hall/CRC, Philadelphia, PA. doi:10.1201/9781584889977.

Bécue-Bertaut, M. (2019). *Textual Data Science with R*. Chapman and Hall/CRC. doi:10.1201/9781315212661.

Bécue-Bertaut, M., Alvarez-Esteban, R., and Canals, J.M. (2003). Use of lexicometry in sensometrics, an essential complement to holistic methods an original methodology. In Springer, ed., *Consumer Research Methods in Food Science. Methods and Protocols in Food Science*, chap. 19, 275–294. Springer, New York. doi:10.1007/978-1-0716-3000-6_19.

Bécue-Bertaut, M., Alvarez-Esteban, R., and Pagès, J. (2008). Rating of products through scores and free-text assertions. comparing and combining both. *Food Quality and Preference*, 19(1): 122–134.

Bécue-Bertaut, M., Alvarez-Esteban, R., Sánchez-Espigares, J.A., and Kostov, B. (2024). *Xplortext: Statistical Analysis of Textual Data.* doi:10.32614/CRAN.package.Xplortext. R package version 1.5.4.

Bécue-Bertaut, M., Kostov, B., Morin, A., and Naro, G. (2014a). Rhetorical strategy in forensic speeches: Multidimensional statistics-based methodology. *J. Classif.*, 31(1): 85–106. doi:10.1007/s00357-014-9148-9.

Bécue-Bertaut, M. and Pagès, J. (2004). A principal axes method for comparing contingency tables: MFACT. *Computational Statistics and Data Analysis*, 45(3): 481–503. doi:10.1016/S0167-9473(03)00003-3.

Bécue-Bertaut, M. and Pagès, J. (2008). Multiple factor analysis and clustering of a mixture of quantitative, categorical and frequency data. *Computational Statistics and Data Analysis*, 52: 3255–3268. doi:10.1016/j.csda.2007.09.023.

Bécue-Bertaut, M. and Pagès, J. (2015). Correspondence analysis of textual data involving contextual information: CA-GALT on principal components. *Advances in Data Analysis and Classification*, 9(2): 125–142. doi:10.1007/s11634-014-0171-9.

Bécue-Bertaut, M., Pagès, J., and Kostov, B. (2014b). Untangling the influence of several contextual variables on the respondents' lexical choices. A statistical approach. *SORT*, 38: 285–302.

Benoit, K., Watanabe, K., Wang, H., Nulty, P., Obeng, A., Müller, S., and Matsuo, A. (2018). quanteda: An R package for the quantitative analysis of textual data. *Journal of Open Source Software*, 3(30): 774. doi:10.21105/joss.00774.

Benzécri, J.P. (1973). *L'Analyse des Données, Tome I Taxinomie, Tome II Correspondances.* Dunod, Paris.

Benzécri, J.P. (1981). *Pratique de l'analyse des données, Tome III, Linguistique and Lexicologie.* Dunod, Paris.

Benzécri, J. (1983). Analyse de l'inertie intraclasse par l'analyse d'un tableau de contingence. *Cahiers de l'analyse des données*, 8: 351–358.

Benzécri, J.P. (1992). *Correspondence Analysis Handbook.* Marcel Dekker, Inc., New York.

Berkhin, P. (2006). A survey of clustering data mining techniques. J. Kogan, C. Nicholas, and M. Teboulle, eds., *Grouping Multidimensional Data*, 25–71. Springer-Verlag, Berlin/Heidelberg.

Bertin, J. (2010). *Semiology of Graphics: Diagrams, Networks, Maps.* ESRI Press, Second Edition, First Edition in French 1979 edn.

Blasius, J. and Greenacre, M., J. (1998). *Visualization of Categorical Data.* Academic Press, San Diego, CA. doi:10.1016/B978-0-12-299045-8.X5000-1.

Blasius, J. and Greenacre, M., J. (2014). *Visualization and Verbalization of Data.* Chapman & Hall/CRC. Computer Science & Data Analysis. Chapman & Hall/CRC, Philadelphia, PA. doi:10.1201/b16741.

Blum, A. and Mitchell, T. (1998). Combining labeled and unlabeled data with co-training. *Proceedings of the eleventh annual conference on Computational learning theory*, 92–100. doi:10.1145/279943.279962.

Borcard, D., Gillet, F., and Legendre, P. (2018). *Numerical Ecology with R.* Use R! Springer International Publishing, Cham, Switzerland, 2 edn.

Brandimarte, P. (2011). *Quantitative Methods: An Introduction for Business Management.* John Wiley & Sons.

Cao, N. and Cui, W. (2016). *Introduction to Text Visualization.* Atlantis Briefs in Artificial Intelligence. Atlantis Press (Zeger Karssen), Amsterdam, Netherlands, 1 edn. doi:10.2991/978-94-6239-186-4.

Cariou, V. and Qannari, E.M. (2018). Statistical treatment of free sorting data by means of correspondence and cluster analyses. *Food Quality and Preference*, 68: 1–11. doi:https://doi.org/10.1016/j.foodqual.2018.01.011.

Cazes, P. and Moreau, J. (2010). Analysis of a contingency table in which the rows and the columns have a graph structure. L.Y. Diday E., ed., *Symbolic-Numeric Data Analysis and Learning*, 271–280. Nova Science Publishers, New York.

Chavent, M., Kuentz-Simonet, V., Labenne, A., and Saracco, J. (2015). *ClustGeo: Hierarchical Clustering with Spatial Constraints.* doi:10.32614/CRAN.package.ClustGeo.

Chavent, M., Kuentz-Simonet, V., Labenne, A., and Saracco, J. (2018). Clustgeo: an R package for hierarchical clustering with spatial constraints. *Computational Statistics*, 33(4): 1799–1882. doi:10.1007/s00180-018-0791-1.

Deerwester, S., Dumais, S.T., Furnas, G.W., Landauer, T.K., and Harshman, R. (1990). Indexing by latent semantic analysis. *J. Am. Soc. Inf. Sci.*, 41(6): 391–407.

Desgraupes, B. (2023). *clusterCrit: Clustering Indices.* doi:10.32614/CRAN.package.clusterCrit. R package version 1.3.0.

Devlin, J., Chang, M.W., Lee, K., and Toutanova, K. (2019). Bert: Pre-training of deep bidirectional transformers for language understanding. doi:10.48550/arXiv.1810.04805.

Efron, B. (1979). Bootstrap methods: Another look at the jacknife. *Ann. Statist.*, 7: 1–26.

Escofier, B. and Drouet, D. (1983). Analyse des différences entre plusieurs tableaux de fréquence. *Cahiers de l'analyse des données*, 8: 491–499.

Escofier, B. and Pagès, J. (2016). *Analyses factorielles simples et multiples.* Dunod, 5th edn.

Escofier [Cordier], B. (1965). *Analyse factorielle des correspondances.* Ph.D. thesis, Faculté des Sciences de Rennes, Rennes.

Ester, M., Kriegel, H.P., Sander, J., and Xu, X. (1996). A density-based algorithm for discovering clusters in large spatial databases with noise. *Proceedings of the Second International Conference on Knowledge Discovery and Data Mining*, KDD'96, 226–231. AAAI Press. doi:10.5555/3001460.3001507.

Feinerer, I. and Hornik, K. (2025). *tm: Text Mining Package.* doi:10.32614/CRAN.package.tm. R package version 0.7-16.

Ferligoj, A. and Batagelj, V. (1982). Clustering with relational constraint. *Psychometrika*, 47(4): 413–426. doi:10.1007/BF02293706.

Friendly, M., Meyer, D., and Zeileis, A. (2015). *Discrete Data Analysis with R: Visualization and Modeling Techniques for Categorical and Count Data.* Chapman and Hall/CRC. doi:10.1201/b19022.

Friendly, M. and Wainer, H. (2021). *A History of Data Visualization and Graphic Communication.* Harvard University Press. doi:10.4159/9780674259034.

Gançarski, P., Dao, T.B.H., Crémilleux, B., Forestier, G., and Lampert, T. (2020). Constrained clustering: Current and new trends. *A Guided Tour of Artificial Intelligence Research*, 447–484. Springer International Publishing, Cham. doi:10.1007/978-3-030-06167-8_14.

Gesmann, M. and de Castillo, D. (2011). googleVis: Interface between R and the google visualisation api. *The R Journal*, 3(2): 40–44.

Glasbey, C.A. (1987). Complete linkage as a multiple stopping rule for single linkage clustering. *J. Classif.*, 4(1): 103–109. doi:10.1007/BF01890078.

González-Almagro, G., Peralta, D., De Poorter, E., Cano, J.R., and García, S. (2025). Semi-supervised constrained clustering: an in-depth overview, ranked taxonomy and future research directions. *Artif. Intell. Rev.*, 58(5). doi:10.1007/s10462-024-11103-8.

Gordon, A. (1999). *Classification.* Chapman and Hall/CRC.

Hahsler, M., Piekenbrock, M., and Doran, D. (2019). dbscan: Fast density-based clustering with R. *Journal of Statistical Software*, 91(1). doi:10.18637/jss.v091.i01.

Handler, A., Denny, M., Wallach, H., and O'Connor, B. (2016). Bag of what? simple noun phrase extraction for text analysis. *Proceedings of the First Workshop on NLP and Computational Social Science*, 114–124. Association for Computational Linguistics. doi:10.18653/v1/w16-5615.

Härdle, W. and L.Simar (2012). *Applied Multivariate Statistical Analysis.* Springer Verlag, Heidelberg.

Harris, Z. (1954). Distributional structure. *Word*, 10(23): 146–162.

Hennig, C., Meila, M., Murtagh, F., and Rocci, R., eds. (2015). *Handbook of Cluster Analysis.* Chapman and Hall/CRC. doi:10.1201/b19706.

Hiep, T.K., Duc, N.M., and Trung, B.Q. (2016a). conclust: Pairwise constraints clustering. doi:10.32614/cran.package.conclust. R package version 1.1.

Hiep, T.K., Duc, N.M., and Trung, B.Q. (2016b). Local search approach for the pairwise constrained clustering problem. SoICT '16, 115–122. Association for Computing Machinery, New York, NY, USA. doi:10.1145/3011077.3011114.

Higgins, M.J., Sävje, F., and Sekhon, J.S. (2016). Improving massive experiments with threshold blocking. *Proceedings of the National Academy of Sciences*, 113(27): 7369–7376. doi:10.1073/pnas.1510504113.

Hill, M.O. (1974). Correspondence analysis: A neglected multivariate method. *J. R. Stat. Soc. Ser. C. Appl. Stat.*, 23(3): 340–354.

Hill, M.O. and Gauch, H.G. (1980). Detrended correspondence analysis: An improved ordination technique. *Vegetatio*, 42(1–3): 47–58. doi:10.1007/bf00048870.

Hofmann, M. and Chisholm, A., eds. (2020). *Text mining and visualization.* Chapman & Hall/CRC Data Mining and Knowledge Discovery Series. Chapman & Hall/CRC, Philadelphia, PA. doi:10.1201/b19007.

Hornik, K., Meyer, D., and Buchta, C. (2009). slam: Sparse lightweight arrays and matrices. doi:10.32614/CRAN.package.slam. Title of the publication associated with this dataset: CRAN: Contributed Packages.

Hubert, L. and Arabie, P. (1985). Comparing partitions. *J. Classif.*, 2(1): 193–218.

Husson, F., L, S., and Pagès, J. (2009). *Analyse de données avec R.* Presse universitaire de Rennes, Rennes.

Jain, A.K. (2010). Data clustering: 50 years beyond k-means. *Pattern Recognit. Lett.*, 31(8): 651–666.

Josse, J., Pages, J., and Husson, F. (2008). Testing the significance of the RV coefficient. *Computational Statistics and Data Analysis*, 53: 82–91.

Kostov, B. (2015). *A principal component method to analyse disconnected frequency tables by means of contextual information.* Ph.D. thesis, UPC, Departament d'Estadística i Investigación Operativa.

Kostov, B., Alvarez-Esteban, R., Bécue-Bertaut, M., and Husson, F. (2024). Multilingual textual data. an approach through multiple factor analysis. *Italian Journal of Applied Statistics*, 35(3): 339–357.

Kostov, B., Bécue-Bertaut, M., and Husson, F. (2015). Correspondence analysis on generalised aggregated lexical tables (ca-galt) in the FactoMineR package. *The R Journal*, 7: 109–117. doi:10.32614/RJ-2015-010.

Kucher, K. and Kerren, A. (2015). Text visualization techniques: Taxonomy, visual survey, and community insights. *2015 IEEE Pacific Visualization Symposium (PacificVis).* IEEE. doi:10.1109/PACIFICVIS.2015.7156366.

Lance, G.N. and Williams, W.T. (1967). A general theory of classificatory sorting strategies: 1. Hierarchical systems. *Comput. J.*, 9(4): 373–380.

Lebart, L. (1978). Programme d'agrégation avec contraintes. *Cahiers de l'analyse des données*, 3(3): 275–287.

Lebart, L. (2007). Which bootstrap for principal axes methods? *Selected Contributions in Data Analysis and Classification*, Studies in classification, data analysis, and knowledge organization, 581–588. Springer Berlin Heidelberg, Berlin, Heidelberg. doi:10.1007/978-3-540-73560-1_55.

Lebart, L. and Mirkin, B. (1993). Correspondence analysis and classification. C. CUADRAS and C. RAO, eds., *Multivariate Analysis: Future Directions 2*, North-Holland Series in Statistics and Probability, 341–357. North-Holland, Amsterdam. doi:10.1016/B978-0-444-81531-6.50024-6.

Lebart, L., Pincemin, B., and Poudat, C. (2019). *Analyse des données textuelles.* Presses de l'Université du Québec. doi:10.3917/popu.2004.0630.

Lebart, L., Piron, M., and Morineau, A. (2006). *Statistique Exploratoire Multidimensionnelle.* Dunod, 4th edn.

Lebart, L., Salem, A., and Berry, L. (1998). *Exploring Textual Data.* Kluwer, Dordrecht.

Lebreton, J., Chessel, D., Prodon, R., and Yoccoz, N. (1988a). L'analyse des relations espèces-milieu par l'analyse canonique des correspondances. I. variables de milieu quantitatives. *Acta Œcologica, Œcologia Generalis*, 9(1): 53–67.

Lebreton, J., Chessel, D., Richardot-Coulet, M., and Yoccoz, N. (1988b). L'analyse des relations espèces-milieu par l'analyse canonique des correspondances. i. variables de milieu qualitatives. *Acta Œcologica, Œcologia Generalis*, 9(2): 137–151.

Legendre, P., Dallot, S., and Legendre, L. (1985). Succession of species within a community: Chronological clustering, with applications to marine and freshwater zooplankton. *Am. Nat.*, 125(2): 257–288. doi:10.1086/284340.

Legendre, P. and Legendre, L. (2012). *Numerical Ecology (Developments in Environmental Modelling, Volume 24)*. Elsevier.

Marriott, K. (2024). *The Golden Age of Data Visualization: How Did We Get Here?* CRC Press. doi:10.1201/9781003507642.

McCune, B. and Grace, J. (2002). *Analysis of Ecological Communities*. MJM Software Design.

McInnes, L., Healy, J., and Melville, J. (2018). UMAP: Uniform manifold approximation and projection for dimension reduction. doi:10.48550/arXiv.1802.03426.

Mihalcea, R. and Tarau, P. (2004). TextRank: Bringing order into text. D. Lin and D. Wu, eds., *Proceedings of the 2004 Conference on Empirical Methods in Natural Language Processing*, 404–411. Association for Computational Linguistics, Barcelona, Spain.

Mikolov, T., Sutskever, I., Chen, K., Corrado, G., and Dean, J. (2013). Distributed representations of words and phrases and their compositionality. doi:10.48550/arXiv.1310.4546.

Milligan, G.W. and Cooper, M.C. (1985). An examination of procedures for determining the number of clusters in a data set. *Psychometrika*, 50: 159–179. doi:https://doi.org/10.1007/BF02294245.

Mouselimis, L. (2021). *fastText: Efficient Learning of Word Representations and Sentence Classification using R*. doi:10.32614/CRAN.package.fastText. R package version 1.0.3.

Murtagh, F. (1985). A survey of algorithms for contiguity-constrained clustering and related problems. *Comput. J.*, 28(1): 82–88. doi:10.1093/comjnl/28.1.82.

Murtagh, F. (2005). *Correspondence Analysis and Data Coding with R and Java*. Chapman and Hall/CRC Press, Boca Raton, FL.

Murtagh, F. and Contreras, P. (2012). Algorithms for hierarchical clustering: an overview. *Wiley Interdiscip. Rev. Data Min. Knowl. Discov.*, 2(1): 86–97. doi:10.1093/comjnl/9.4.373.

Nielsen, F. (2016). Hierarchical clustering. *Introduction to HPC with MPI for Data Science*, 195–211. Springer International Publishing, Cham. doi: 10.1007/978-3-319-21903-5_8.

Nigam, K., McCallum, A., Thrun, S., and Mitchell, T.M. (2000). Text classification from labeled and unlabeled documents using em. *Machine Learning*, 39: 103–134.

Okabe, A., Boots, B., Sugihara, K., Chiu, S.N., and Kendall, D.G. (2000). *Spatial Tessellations: Concepts and Applications of Voronoi Diagrams, Second Edition.* Wiley Series in Probability and Mathematical Statistics. Wiley. doi:10.1002/9780470317013.

Økland, R.H. (1990). Vegetation ecology: theory, methods and applications with reference to fennoscandia. *Sommerfeltia*, 1(s1): 1–238. doi:10.2478/som-1990-0003.

Oksanen, J. and Minchin, P.R. (1997). Instability of ordination results under changes in input data order: explanations and remedies. *J. Veg. Sci.*, 8(3): 447–454. doi:https://doi.org/10.2307/3237336.

Owsiński, J.W., Stańczak, J., Opara, K., Zadrożny, S., and Kacprzyk, J. (2021). *Reverse clustering.* Studies in computational intelligence. Springer Nature, Cham, Switzerland, 2021 edn. doi:10.1007/978-3-030-69359-6.

Pagès, J. (2014). *Multiple Factor Analysis by Example Using R.* Chapman and Hall/CRC.

Pebesma, E. (2016). *sf: Simple Features for R.* doi:10.32614/CRAN.package.sf. R package version 1.0-21.

Qi, Q., Hessen, D., Deoskar, T., and van der Heijden, P. (2024). A comparison of latent semantic analysis and correspondence analysis of document-term matrices. *Natural Language Engineering*, 30: 722–752. doi:10.1017/S1351324923000244.

R Core Team (2025). *R: A Language and Environment for Statistical Computing.* R Foundation for Statistical Computing, Vienna, Austria. R version 4.5.0.

Rand, W.M. (1971). Objective criteria for the evaluation of clustering methods. *J. Am. Stat. Assoc.*, 66(336): 846–850.

Reinert, A. (1983). Une méthode de classification descendante hiérarchique: application à l'analyse lexicale par contexte. *Cahiers de l'analyse des données*, 8(2): 187–198.

Rinker, T.W. (2023). *qdap: Quantitative Discourse Analysis Package.* Buffalo, New York. doi:10.32614/CRAN.package.qdap. 2.4.6.

Rose, S., Engel, D., Cramer, N., and Cowley, W. (2010). Automatic keyword extraction from individual documents. *Text Mining*, 1–20. John Wiley & Sons, Ltd, Chichester, UK. doi:10.1002/9780470689646.ch1.

Rousseeuw, P.J. (1987). Silhouettes: A graphical aid to the interpretation and validation of cluster analysis. *J. Comput. Appl. Math.*, 20: 53–65. doi: 10.1016/0377-0427(87)90125-7.

Salem, A. (1987). *Pratique des segments répétés. Essai de statistique textuelle.* Klincksieck, Paris.

Savje, F., Higgins, M., and Sekhon, J. (2024). *scclust: Size-Constrained Clustering.* doi:10.32614/CRAN.package.scclust. R package version 0.2.5.

Scrucca, L., Fraley, C., Murphy, T.B., and Raftery, A.E. (2023). *Model-Based Clustering, Classification, and Density Estimation Using mclust in R.* Chapman and Hall/CRC. doi:10.1201/9781003277965.

Selivanov, D., Bickel, M., and Wang, Q. (2016). text2vec: Modern text mining framework for R. doi:10.32614/cran.package.text2vec.

Sievert, C. (2020). *Interactive Web-Based Data Visualization with R, plotly, and shiny.* Chapman and Hall/CRC. doi:10.1201/9780429447273.

Studer, M. (2013). Weightedcluster library manual: A practical guide to creating typologies of trajectories in the social sciences with R. *LIVES Working papers*, 24: 1–32. doi:10.12682/lives.2296-1658.2013.24. Place: Lausanne Publisher: NCCR LIVES.

Takane, Y., Yanai, H., and Mayekawa, S. (1991). Relationships among several methods of linearly constrained correspondence analysis. *Psychometrika*, 56: 667–684. doi:10.1007/BF02294498.

ter Braak, C.J.F. (1986). Canonical correspondence analysis: a new eigenvector technique for multivariate direct gradient analysis. *Ecology*, 67: 1167–1179.

ter Braak, C. (1987). *Canoco–a FORTRAN program for canonical community ordination by [partial] [detrended] [canonical] correspondence analysis (version 2.1).* ITI-TNO Institute of Applied Computer Sciences, Wageningen.

Valencia-García, O. and Alvarez-Esteban, R. (2012). A bootstrap procedure in the context of correspondence analysis: Numerical approach to measure the stability of axes. *Studies of Applied Economics*, 30(1): 340–354. doi: 10.25115/eea.v39i9.6996.

van der Maaten, L. and Hinton, G. (2008). Visualizing data using t-sne. *J. Mach. Learn. Res.*, 9(86): 2579–2605.

Vendramin, L., Campello, R.J.G.B., and Hruschka, E.R. (2010). Relative clustering validity criteria: A comparative overview. *Statistical Analysis and Data Mining: The ASA Data Science Journal*, 3.

von Luxburg, U., Williamson, R.C., and Guyon, I. (2012). Clustering: Science or art? I. Guyon, G. Dror, V. Lemaire, G. Taylor, and D. Silver, eds., *Proceedings of ICML Workshop on Unsupervised and Transfer Learning*, vol. 27 of *Proceedings of Machine Learning Research*, 65–79. PMLR, Bellevue, Washington, USA.

Ward, J.H. (1963). Hierarchical grouping to optimize an objective function. *Journal of the American Statistical Association*, 58: 236–244. doi:10.1080/01621459.1963.10500845.

Wijffels, J. (2017). Textrank: Summarize text by ranking sentences and finding keywords. doi:10.32614/CRAN.package.textrank.

Xie, Y. (2015). *Dynamic Documents with R and knitr, Second Edition.* Chapman and Hall/CRC. doi:10.1201/9781315382487.

Xie, Y., Allaire, J.J., and Grolemund, G. (2018). *R markdown.* Chapman and Hall/CRC. doi:10.1201/9781138359444.

Xu, R. and Wunsch, D. (2005). Survey of clustering algorithms. *IEEE Transactions on Neural Networks*, 16(3): 645–678. doi:10.1109/TNN.2005.845141.

Index

For Product Safety Concerns and Information please contact our EU
representative GPSR@taylorandfrancis.com
Taylor & Francis Verlag GmbH, Kaufingerstraße 24, 80331 München, Germany

www.ingramcontent.com/pod-product-compliance
Ingram Content Group UK Ltd.
Pitfield, Milton Keynes, MK11 3LW, UK
UKHW020158200926
475702UK00004B/515
* 9 7 8 1 0 3 2 5 0 2 2 6 7 *